Titanium Alloys - Novel Aspects of Their Manufacturing and Processing

Edited by Maciej Motyka, Waldemar Ziaja and Jan Sieniawski

Published in London, United Kingdom

IntechOpen

Supporting open minds since 2005

Titanium Alloys - Novel Aspects of Their Manufacturing and Processing
http://dx.doi.org/10.5772/intechopen.78626
Edited by Maciej Motyka, Waldemar Ziaja and Jan Sieniawski

Contributors
Imran Masood, Agripa Hamweendo, Ione Botef, Vasyl Trush, Kadephi Vuyolwethu Mjali, Annelize Botes, Mikhail Radkevich, Chakradhar Bandapalli, Bharatkumar Mohanbhai Sutaria, Dhananjay Vishnu Prasad Bhatt, Nageswara Rao Muktinutalapati, Sudhagara Rajan S, Jithin Vishnu, Geetha Manivasagam, Maciej Motyka

First published in London, United Kingdom, 2019 by IntechOpen
IntechOpen is the global imprint of INTECHOPEN LIMITED, registered in England and Wales, registration number: 11086078, 7th floor, 10 Lower Thames Street, London, EC3R 6AF, United Kingdom
Printed in Croatia

British Library Cataloguing-in-Publication Data
A catalogue record for this book is available from the British Library

Additional hard and PDF copies can be obtained from orders@intechopen.com

Titanium Alloys - Novel Aspects of Their Manufacturing and Processing
Edited by Maciej Motyka, Waldemar Ziaja and Jan Sieniawski
p. cm.
Print ISBN 978-1-83962-552-7
Online ISBN 978-1-83962-553-4
eBook (PDF) ISBN 978-1-83962-554-1

For EU product safety concerns:
IN TECH d.o.o., Prolaz Marije Krucifikse Kozulić 3, 51000 Rijeka, Croatia, info@intechopen.com or visit our website at intechopen.com.

Meet the editors

Maciej Motyka serves as an Associate Professor in the Department of Materials Science at the Faculty of Mechanical Engineering and Aeronautics of the Rzeszow University of Technology. His scientific interests cover relationships between processing, microstructure, and mechanical properties of advanced structural materials, mainly titanium alloys. His activity is focused on hot plasticity and fine-grained superplasticity phenomena. He is also working on characterization of ultrafine-grained materials – submicrocrystalline aluminum alloys and nanocrystalline titanium – obtained by plastic consolidation and severe plastic deformation methods.

Waldemar Ziaja is engaged with the Department of Material Science at the Faculty of Mechanical Engineering and Aeronautics of the Rzeszow University of Technology. His scientific interests include relationships between processing, microstructure, and mechanical properties of advanced structural materials, mainly titanium and nickel-based alloys. He is also working on modelling the microstructure-dependent deformation of materials in fatigue and creep conditions including damage initiation and growth.

Professor Jan Sieniawski holds the management position in the Department of Material Science at the Faculty of Mechanical Engineering and Aeronautics and R&D Laboratory of Aerospace Materials of the Rzeszow University of Technology. His scientific activity covers topics in the area of physical metallurgy of aerospace materials – titanium and aluminum alloys, nickel- and cobalt-based superalloys, development and characterization of single crystal castings, and thermal barrier coatings on the elements of the hot section of gas turbines.
Prof. Sieniawski is a founder member of the Polish Materials Society and a member of The Committee on Materials Science and Committee on Machine Building of the Polish Academy of Sciences.

Contents

Preface

Titanium alloys, due to unique physical and chemical properties (mainly high relative strength combined with very good corrosion resistance), are considered as an important structural metallic material used in hi-tech industries (e.g. aerospace, space technology). Their development has led to the design of several groups of structural alloys, including single-phase α or β alloys, two-phase α+β alloys, and TiAl intermetallic alloys. Undoubtedly the consumption of titanium alloys has been continuously increasing in recent years. Although titanium and its alloys have been extensively researched for over 50 years, their application potential is still high. This has been confirmed by numerous recent publications and books presenting new findings on novel application areas for titanium alloys. This book aims to provide information on new processing methods of single- and two-phase titanium alloys.

The eight chapters of this book are distributed over four sections. The first section (*Introduction*) indicates the main factors determining application areas of titanium and its alloys. In the introductory chapter, the recent trends in design of titanium alloys and advanced technologies used for their processing are described briefly. The second section (*Manufacturing*, two chapters) concerns modern production methods for titanium and its alloys.

The next section (*Thermomechanical and surface treatment*, three chapters) covers problems of thermomechanical processing and surface treatment used for single- and two-phase titanium alloys. The fourth section (*Machining*, two chapters) describes recent results of high-speed machining of Ti-6Al-4V alloy and the possibility of application of sustainable machining for titanium alloys.

The first chapter is the introduction. The second chapter reviews the modern production methods for titanium alloys. It presents the current methods used for modern applications. The author also discuss the future development related to the most probable demands of titanium and titanium alloy products.

The third chapter is devoted to two manufacturing processes intended for commercially pure titanium: laser and mechanical forming. The chapter discusses the most important aspects related to microstructure and mechanical properties of the material and the level of residual stress in the elements after forming.

In the fourth chapter, bulk processing, including vacuum melting and hot working operations, is discussed. The sub-solvus forging is demonstrated as a method resulting in a superior combination of mechanical properties of beta titanium alloys. The chapter attempts to review the studies on manufacturing, plastic working, heat treatment, and surface modification of beta titanium alloys intended for aerospace and biomedical applications.

The fifth chapter describes the plastic flow behaviour of pseudo-alpha titanium alloy deformed at various temperatures and strain rates. The chapter presents processing maps elaborated on the basis of energy dissipation. The possibility of superplastic deformation of the examined alloy is confirmed.

The sixth chapter refers to solid-solution hardening of surface layers of titanium alloys (α, near-α, α+β) due to diffusional saturation in a gas medium containing oxygen. The authors have determined the relationship between parameters of surface-hardened layers (surface hardness, depth of hardened zone, microstructure) obtained by various surface hardening methods and fatigue properties of examined titanium alloys.

The seventh chapter deals with the high-speed micro-milling process used to achieve the desired surface finish without traditional coolants. The investigation of various tool (uncoated & PVD coated AlTiN, TiAlN tungsten carbide) wear behaviour in the milling process under dry cutting conditions of Ti-6Al-4V ELI alloy is presented. Analysis of cutting force and wear mechanisms is performed for examined mills.

In the eighth chapter, an analysis of the machining process of Ti-6Al-4V alloy, both dry and using conventional and cryogenic cooling, is presented. The assessment of machining sustainability is based on the following variables: cutting power, machining time, machining cost, material removal rate, and cutting tool life.

We hope that the recent research data and reviews presented in this book will contribute to the improvement of operational properties and the increase of the range of applications of titanium and its alloys.

D.Sc. Maciej Motyka, Ph.D. Waldemar Ziaja and Prof. Jan Sieniawski
Department of Materials Science,
Faculty of Mechanical Engineering and Aeronautics
of the Rzeszow University of Technology,
Rzeszow, Poland

Section 1

Introduction

Chapter 1

Introductory Chapter: Novel Aspects of Titanium Alloys' Applications

Maciej Motyka, Waldemar Ziaja and Jan Sieniawski

1. Introduction

Titanium is characterized by unique physical and chemical properties determining its specific applications. Since it was discovered in 1791 by William Gregor, its production was considered difficult and unprofitable for almost 150 years. In 1940, William J. Kroll developed commercially attractive production method based on the reduction of $TiCl_4$ using Na or Mg. Kroll process, in substantially unchanged form, is still the dominant process for titanium production. Titanium sponge is remelted (e.g., in vacuum arc process—VAR) to the form of commercial pure (CP) titanium or titanium alloys. Ingots are usually primarily processed by homogenization annealing or plastic working in the β-phase field. Products can be manufactured by casting and plastic working processes [1–5].

Titanium alloys—comparing with other structural materials—are characterized by high relative strength in the wide temperature range and very good corrosion resistance in many chemically aggressive environments. Such properties create many possibilities of improvements of technological processes, tooling and products in various industry branches. The main application areas of titanium alloys include transportation (mainly aerospace industry), chemical, food, machine building, papermaking, electrotechnics, electronic, fuel-energetic, metallurgical industries, and also geology and medicine [6]. Mechanical properties of titanium alloys are developed in plastic working and heat treatment processes, causing intentional microstructure evolution. It should be pointed that obtaining finished products having desired microstructure and properties is difficult due to some of the properties of titanium alloys, such as: high chemical affinity to oxygen, low thermal conductivity, high heat capacity and significant dependence of plastic flow resistance on strain rate. Quite often, hot-worked titanium products are characterized by various deformation conditions leading to formation of zones having various phase composition and dispersion and therefore various mechanical properties [7].

The main types of microstructure in two-phase titanium alloys are lamellar—consisting of colonies of α-phase lamellae within β-phase grains of several hundred microns in diameter (formed after slow cooling when deformation or heat treatment takes place at a temperature above the β-transus)—and equiaxed—consisting of globular α-phase dispersed in β-phase matrix (formed after deformation in the two-phase $\alpha + \beta$ field). Alloys having lamellar microstructure are characterized by relatively low tensile ductility, moderate fatigue properties and good creep and fatigue crack growth resistance, whereas in case of equiaxed microstructure, materials have a better balance of strength and ductility at room temperature and fatigue properties [8].

IntechOpen

2. Perspective titanium alloys

Due to the high chemical affinity of titanium to atmospheric gases, the application of conventional titanium alloys at elevated temperature is limited. Single-phase α alloys can be used up to 600°C. Much higher heat resistance is achieved in intermetallic TiAl(gamma)-based alloys. They exhibit superior specific strength-temperature properties, comparing to classical titanium alloys, steels, and nickel-based superalloys, in the temperature range from 500 to 900°C. Nowadays, TiAl-based alloys are considered as a high-potential material for aircraft engines. The main problems related to their applications are low ductility and the difficulty in processing to form a component. Over the last 30 years, three generations of gamma aluminide titanium alloys have been developed and the basic concept of the fourth generation has been described [9].

Another important feature of titanium alloys is the low value of Young's modulus (E)—from about 100 (for β-phase alloys) to 125 GPa (for α-phase alloys). Low-Young's modulus titanium alloys are considered as valuable biomaterials used for bone implants. It allows to prevent stress shielding, which usually leads to bone resorption and poor bone remodeling, when metal implants are used. The new generation of β-type titanium alloys composed of nontoxic and allergy-free elements, the so-called TNTZ alloys (e.g., Ti–29Nb–13Ta–4.6Zr), is characterized by Young's modulus lower than 90 GPa [10, 11]. TNTZ alloys can exhibit the E value lower than 20 GPa—they are called "gum metals." It was found that the most important role in terms of obtaining the outstanding mechanical properties and the unique deformation behavior plays the oxygen content (stabilizes the bcc crystal structure by controlling the martensitic transformation temperature) [12].

Some of the β-type titanium alloys seem to have potential for even broader range of application due to shape memory effect. Shape memory alloys (SMA), especially Ni-Ti alloys (Nitinols), in recent years, have been mainly applied for biomedical implants and devices. However, due to the risk of Ni allergy and hypersensitivity, their long-term use is limited. The β-type Ti-based SMA have been extensively studied as promising candidates for Ni-free biomedical shape memory alloys [13].

3. Advanced material technologies

Novel aspects of material applications are also related to modern manufacturing and processing technologies. It is worth to note about the grain refinement, which causes high strength increase in metallic materials. Severe plastic deformation (SPD) methods allow to achieve high mechanical properties in conventional titanium alloys. Pure nanocrystalline titanium is characterized by the strength level very close to solution-strengthened titanium alloys (e.g., Ti-6Al-4V) [14]. Ultrafine-grained titanium alloys exhibit high superplastic deformability. Superplastic forming combining with diffusion bonding (SPF/DB) is a well-established method used in aerospace industry for the production of complex-shaped sheet elements made of titanium alloys [15].

Other developing processing areas are: surface engineering, joining methods (e.g., friction stir welding—FSW), and machining [3, 4]. Highly promising are especially additive manufacturing (AM) methods. In contrast to conventional processes (casting, plastic working, or expensive machining), the AM allows to produce near-net shape structural parts—minimizing finishing techniques cost (machining) and achieving mechanical properties at least at the level of cast and wrought products. AM of titanium alloys has been used quite quickly to many applications in aerospace and medical industry. Titanium and its alloys are considered as ideal materials for the additive manufacturing industry [16].

Author details

Maciej Motyka*, Waldemar Ziaja and Jan Sieniawski
Department of Materials Science, Faculty of Mechanical Engineering and Aeronautics, Rzeszow University of Technology, Rzeszow, Poland

*Address all correspondence to: motyka@prz.edu.pl

References

[1] Zwicker U. Titanium and Titanium Alloys (in German). Berlin-Heidelberg-New York: Springer Verlag; 1974

[2] Bylica A, Sieniawski J. Titanium and Its Alloys (in Polish). Warsaw: PWN; 1985

[3] Lütjering G, Williams JC. Titanium. Berlin-Heidelberg-New York: Springer Verlag; 2003

[4] Donachie MJ. Titanium. A Technical Guide. Materials Park OH: ASM International; 2000

[5] Banerjee D, Williams JC. Perspective on titanium science and technology. Acta Materialia. 2013;**61**:844-879

[6] Motyka M, Kubiak K, Sieniawski J, Ziaja W. Phase transformations and characterization of α+β titanium alloys. In: Hashmi S, editor. Comprehensive Materials Processing–Volume 2: Materials Modeling and Characterization. Amsterdam: Elsevier; 2014

[7] Motyka M, Kubiak K, Sieniawski J, Ziaja W. Hot plasticity of alpha beta alloys. In: Nurul Amin AKM, editor. Titanium Alloys—Towards Achieving Enhanced Properties for Diversified Applications. Rijeka: InTech; 2012

[8] Sieniawski J, Ziaja W, Kubiak K, Motyka M. Microstructure and mechanical properties of high strength two-phase titanium alloys. In: Sieniawski J, Ziaja W, editors. Titanium Alloys—Advances in Properties Control. Rijeka: InTech; 2013

[9] Lapin J. TiAl-based alloys. Present status and future perspectives. In: Proceedings of 18th International Metallurgical & Materials Conference Proceedings METAL 2009. 19-21 May 2009, Hradec nad Moravicí, Czech Republic

[10] Khorasani AM, Goldberg M, Doeven EH, Littlefair G. Titanium in biomedical applications—Properties and fabrication: A review. Journal of Biomaterials and Tissue Engineering. 2015;**5**:593-619

[11] Niinomi M, Liu Y, Nakai M, Liu H, Li H. Biomedical titanium alloys with Young's moduli close to that of cortical bone. Regenerative Biomaterials. 2016;**3**(3):173-185

[12] Furuta T, Kuramoto S, Morris JW, Nagasako N Jr, Withey E, Chrzan DC. The mechanism of strength and deformation in gum metal. Scripta Materialia. 2013;**68**:767-773

[13] Kim HY, Miyazaki S. Ni-Free Ti-Based Shape Memory Alloys. Oxford: Butterworth-Heinemann; 2018

[14] Garbacz H, Semenova IP, Zherebtsov S, Motyka M. Nanocrystalline Titanium. Amsterdam: Elsevier; 2019

[15] Sieniawski J, Motyka M. Superplasticity in titanium alloys. Journal of Achievements in Materials and Manufacturing Engineering. 2007;**24**(1):123-130

[16] Dutta B, Froes FH. Additive Manufacturing of Titanium Alloys. Oxford: Butterworth-Heinemann; 2016

Section 2

Manufacturing

Chapter 2

Modern Production Methods for Titanium Alloys: A Review

Hamweendo Agripa and Ionel Botef

Abstract

Titanium alloys are advanced structural materials for numerous key engineering applications in medicine (implants), aerospace, marine structures, and many other areas. The novel aspects of application potential for titanium alloys are as a result of their unique properties such as high corrosion resistance, high specific strength, low elastic modulus, high elasticity, and high hardness. This chapter examines the modern methods for production of titanium alloys. The goal of this chapter is to show the process engineers the current methods for production of titanium alloys necessary for modern applications. The chapter also presents the future methods of production for titanium and titanium alloys to meet the future demands of titanium and titanium alloys' products.

Keywords: titanium, titanium alloys, production methods

1. Introduction

Titanium (Ti) is a lustrous metal with a silver color. This metal exists in two different physical crystalline state called body centered cubic (bcc) and hexagonal closed packing (hcp), shown in **Figure 1 (a)** and **(b)**, respectively. Titanium has five natural isotopes, and these are 46Ti, 47Ti, 48Ti, 49Ti, 50Ti. The 48Ti is the most abundant (73.8%).

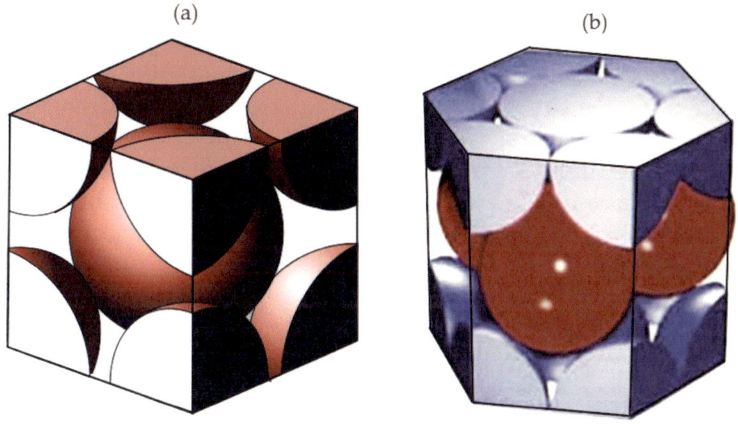

Figure 1.
Crystalline state of titanium: (a) bcc, and (b) hcp [8].

Titanium has high strength of 430 MPa and low density of 4.5 g/cm^3, compared to iron with strength of 200 MPa and density of 7.9 g/cm^3. Accordingly, titanium has the highest strength-to-density ratio than all other metals. However, titanium is quite ductile especially in an oxygen-free environment. In addition, titanium has relatively high melting point (more than 1650°C or 3000°F), and is paramagnetic with fairly low electrical and thermal conductivity. Further, titanium has very low bio-toxicity and is therefore bio-compatible. Furthermore, titanium readily reacts with oxygen at 1200°C (2190°F) in air, and at 610°C (1130°F) in pure oxygen, forming titanium dioxide. At ambient temperature, titanium slowly reacts with water and air to form a passive oxide coating that protects the bulk metal from further oxidation, hence, it has excellent resistance to corrosion and attack by dilute sulfuric and hydrochloric acids, chloride solutions, and most organic acids. However, titanium reacts with pure nitrogen gas at 800°C (1470°F) to form titanium nitride [1, 2].

Some of the major areas where titanium is used include the aerospace industry, orthopedics, dental implants, medical equipment, power generation, nuclear waste storage, automotive components, and food and pharmaceutical manufacturing.

Titanium is the ninth-most abundant element in Earth's crust (0.63% by mass) and the seventh-most abundant metal. The fact that titanium has most useful properties makes it be preferred material of future engineering application. Moreover, the application of titanium can be extended when alloyed with other elements as described below.

2. Titanium alloys

An alloy is a substance composed of two or more elements (metals or nonmetals) that are intimately mixed by fusion or electro-deposition. On this basis, titanium alloys are made by adding elements such as aluminum, vanadium, molybdenum, niobium, zirconium and many others to produce alloys such as Ti-6Al-4V and Ti-24Nb-4Zr-8Sn and several others [2]. These alloys have exceptional properties as illustrated below. Depending on their influence on the heat treating temperature and the alloying elements, the alloys of titanium can be classified into the following three types:

2.1 Type 1: the alpha (α) alloys

These alloys contain a large amount of α-stabilizing alloying elements such as aluminum, oxygen, nitrogen or carbon. Aluminum is widely used as the alpha stabilizer for most commercial titanium alloys because it is capable strengthening the alloy at ambient and elevated temperatures up to about 550°C. This capability coupled with its low density makes aluminum to have additional advantage over other alloying elements such as copper and molybdenum. However, the amount of aluminum that can be added is limited because of the formation of a brittle titanium-aluminum compound when 8% or more by weight aluminum is added. Occasionally, oxygen is added to pure titanium to produce a range of grades having increasing strength as the oxygen level is raised. The limitation of the α alloys of titanium is non-heat treatable but these are generally very weldable. In addition, these alloys have low to medium strength, good notch toughness, reasonably good ductility and have excellent properties at cryogenic temperatures. These alloys can be strengthened further by the addition of tin or zirconium. These metals have appreciable solubility in both alpha and beta phases and as their addition does not markedly influence the transformation temperature they are normally classified as neutral additions. Just like aluminum, the benefit of hardening at ambient temperature is retained even at elevated temperatures when tin and zirconium are used as alloying elements.

2.2 Type 2: alpha-beta (α-β) titanium alloys

These alloys contain 4–6% of β-phase stabilizer elements such as molybdenum, vanadium, tungsten, tantalum, and silicon. The amount of these elements increases the amount of β-phase is the metal matrix. Consequently, these alloys are heat treatable, and are significantly strengthened by precipitation hardening. Solution treatment of these alloys causes increase of β-phase content mechanical strength while ductility decreases. The most popular example of the α-β titanium alloy is the Ti-6Al-4V with 6 and 4% by weight aluminum and vanadium, respectively. This alloy of titanium is about half of all titanium alloys produced. In these alloys, the aluminum is added as α-phase stabilizer and hardener due to its solution strengthening effect. The vanadium stabilizes the ductile β-phase, providing hot workability of the alloy.

2.2.1 Properties of α-β titanium alloys

The α-β titanium alloys have high tensile strength, high fatigue strength, high corrosion resistance, good hot formability and high creep resistance [3].

2.2.2 Novel application of α-β titanium alloys

Therefore, these alloys are used for manufacturing steam turbine blades, gas and chemical pumps, airframes and jet engine parts, pressure vessels, blades and discs of aircraft turbines, aircraft hydraulic tubing, rocket motor cases, cryogenic parts, and marine components [4].

2.3 Type 3: beta (β) titanium alloys

These alloys exhibit the body centered cubic crystalline form shown in **Figure 1 (a)**. The β stabilizing elements used in these alloy are one or more of the following: molybdenum, vanadium, niobium, tantalum, zirconium, manganese, iron, chromium, cobalt, nickel, and copper. Besides strengthening the beta phase, these β stabilizers lower the resistance to deformation which tends to improve alloy fabricability during both hot and cold working operations. In addition, this β stabilizer to titanium compositions also confers a heat treatment capability which permits significant strengthening during the heat treatment process [4].

2.3.1 Properties of beta (β) titanium alloys

As a result, the β titanium alloys have large strength to modulus of elasticity ratios that is almost twice those of 18–8 austenitic stainless steel. In addition, these β titanium alloys contain completely biocompatible elements that impart exceptional biochemical properties such as superior properties such as exceptionally high strength-to-weight ratio, low elastic modulus, super-elasticity low elastic modulus, larger elastic deflections, and low toxicity [1, 3].

2.3.2 Novel application of beta (β) titanium alloys

The above properties make them to be bio-compatible and are excellent prospective materials for manufacturing of bio-implants. Therefore, nowadays these alloys are largely utilized in the orthodontic field since the 1980s, replacing the stainless steel for certain uses, as stainless steel had dominated orthodontics since the 1960s [2].

2.4 Summary

Because of alloying the titanium achieve improved properties that make it to be preferred material of choice for application in aerospace, medical, marine and instrumentation. The extent of improvement to the properties of titanium alloys and ultimately the choice of area of application is influenced by the methods of production and processing as discussed in the subsequent sections.

3. Production of titanium

The base metal required for production of titanium alloys is pure titanium. Pure titanium is produced using several methods including the Kroll process. This process produces the majority of titanium primary metals used globally by industry today. In this process, the titanium is extracted from its ore rutile—TiO_2 or titanium concentrates. These materials are put in a fluidized-bed reactor along with chlorine gas and carbon and heated to 900°C and the subsequent chemical reaction results in the creation of impure titanium tetrachloride ($TiCl_4$) and carbon monoxide. The resultant titanium tetrachloride is fed into vertical distillation tanks where it is heated to remove the impurities by separation using processes such as fractional distillation and precipitation. These processes remove metal chlorides including those of iron, silicon, zirconium, vanadium and magnesium. Thereafter, the purified liquid titanium tetrachloride is transferred to a reactor vessel in which magnesium is added and the container is heated to slightly above 1000°C. At this stage, the argon is pumped into the container to remove the air and prevent the contamination of the titanium with oxygen or nitrogen. During this process, the magnesium reacts with the chlorine to produce liquid magnesium chloride thereby leaving the pure titanium solid. This process is schematically presented in **Figure 2**.

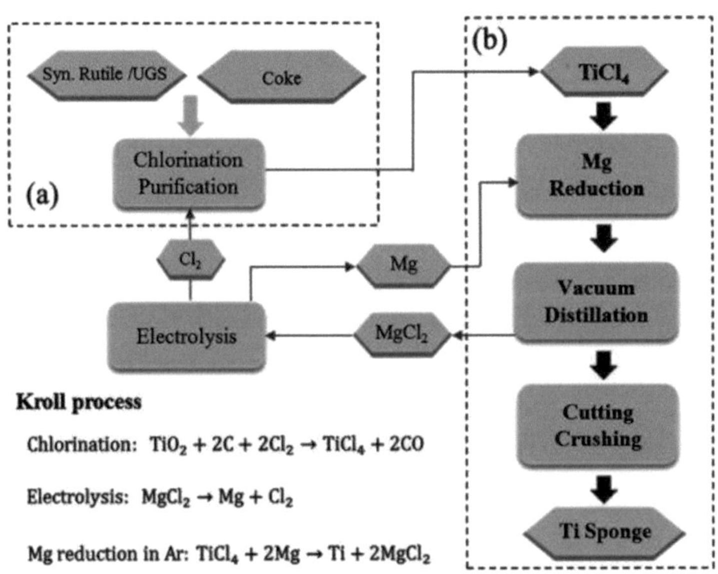

Figure 2.
Kroll process for production of titanium: (a) chlorination, (b) fractional distillation [5].

The resultant titanium solid is removed from the reactor by boring and then treated with water and hydrochloric acid to remove excess magnesium and magnesium chloride leaving porous titanium sponge, which is jackhammered, crushed, and pressed, followed by melting in a vacuum electric arc furnace using expendable carbon electrode. The melted ingot is allowed to solidify in a vacuum atmosphere. This solid is often remelted to remove inclusions and to homogenize its constituents. These melting steps add to the cost of producing titanium, and this cost is usually about six times that of stainless steel. Usually the titanium solid undergo further treatment to produce titanium powder required in alloying process. The basic methods used to produce titanium powder are summarized below.

3.1 Armstrong process

The first method is called the Armstrong process, shown in **Figure 3**, in which the powder is made as the product of extractive processes that produce primary metal powder. This process is capable of producing commercially pure titanium (Ti) powder by the reduction of titanium tetrachloride ($TiCl_4$) and other metal halides using sodium (Na). This process produces powder particles with a unique properties and low bulk density. To improve powder properties such as the particle size distribution and the tap density, additional post processing activities such as dry and wet ball milling are applied. The narrowed particle size distributions are necessary for typical powder metallurgical processes. In addition, the resultant powder's morphology produced by the Armstrong process provide for excellent compressibility and compaction properties that result in dense compacts with increased green strength than those produced by the irregular powders. For this reason, the powders can even be consolidated by traditional powder metallurgy techniques such as uniaxial compaction and cold isostatic pressing. **Figure 4** illustration the scanning electron microscope images of the titanium powders of the

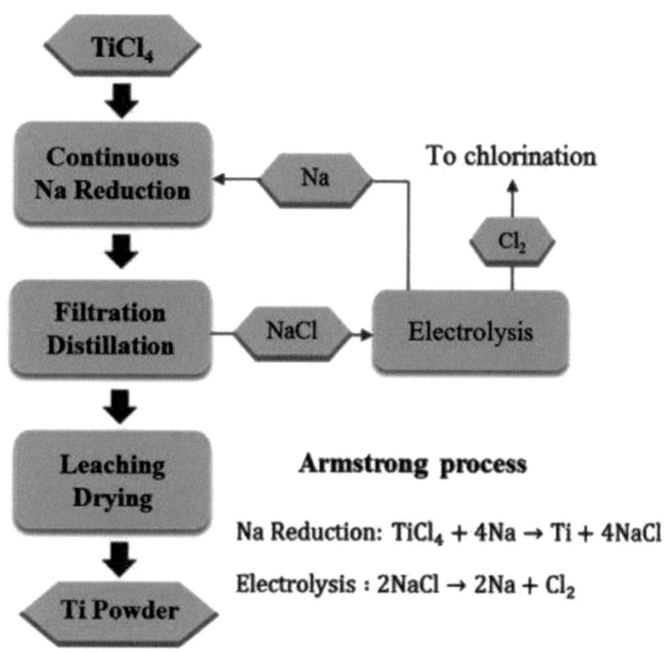

Figure 3.
Illustration of the Armstrong process [5].

Figure 4.
SEM micrographs of CP-Ti produced by Armstrong process [5].

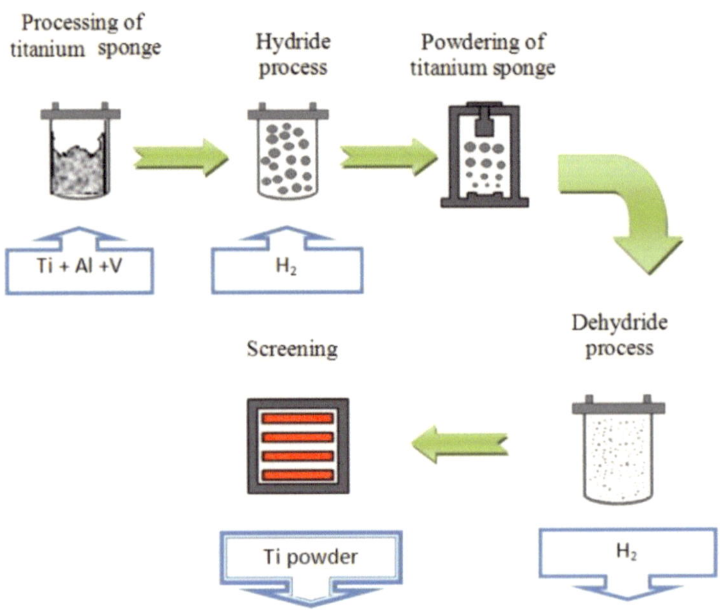

Figure 5.
Hydride-dehydride process for obtaining of titanium powders [6].

Armstrong process. As seen in the figure, the powder has an irregular morphology made of granular agglomerates of smaller particles.

3.2 The hydride-dehydride process

The hydride-dehydride (HDH) process, illustrated in **Figure 5**, is used to produce titanium powder using titanium sponge, titanium, mill products, or titanium scrap as the raw material. The hydrogenation process is achieved using a batch furnace that is usually operated in vacuum and/or hydrogen atmospheric conditions. The conditions necessary for hydrogenation of titanium are pressure of one atmospheric and temperatures of utmost 800°C. This process results in forming of titanium hydride and alloy hydrides that are usually brittle in nature. These metal

Figure 6.
SEM micrographs of CP-Ti produced by HDH [5].

hydrides are milled and screened to produce fine powders. The powder is resized using a variety of powder-crushing and milling techniques may be used including: a jaw crusher, ball milling, or jet milling. After the titanium hydride powders are crushed and classified, they are placed back in the batch furnace to dehydrogenate and remove the interstitial hydrogen under vacuum or argon atmosphere and produce metal powder. These powders are irregular and angular in morphology and can also be magnetically screened and acid washed to remove any ferromagnetic contamination. Finer particle sizes can be obtained, but rarely used because oxygen content increases rapidly when the powder is finer than −325 mesh. Powder finer than −325 mesh also possess more safety challenges [5]. The powder can be passivated upon completion of both the hydrogenating and dehydrogenating cycles to minimize exothermic heat generated when exposed to air.

The hydride-dehydride process is relatively inexpensive because the hydrogenation and dehydrogenation processes contribute small amount of cost to that of input material. The additional benefit of this process is the fact that the purity of the powder can be very high, as long as the raw material's impurities are reduced. The oxygen content of final powder has a strong dependence on the input material, the handling processes and the specific surface area of the powder. Therefore, the main disadvantages of hydride-dehydride powder include: the powder morphology is irregular, and the process is not suitable for making virgin alloyed powders or modification of alloy compositions if the raw material is from scrap alloys (**Figure 6**) [5].

4. Conventional methods of production for titanium alloys

4.1 Powder metallurgy

Conventional sintering, shown in **Figure 7**, is one of the widely applied powder metallurgy (PM) based method for manufacturing titanium alloys. In this method, the feedstock titanium powder is mixed thoroughly with alloying elements mentioned in Section 2 using a suitable powder blender, followed by compaction of the mixture under high pressure, and finally sintered. The sintering operation is carried out at high temperature and pressure treatment process that causes the powder

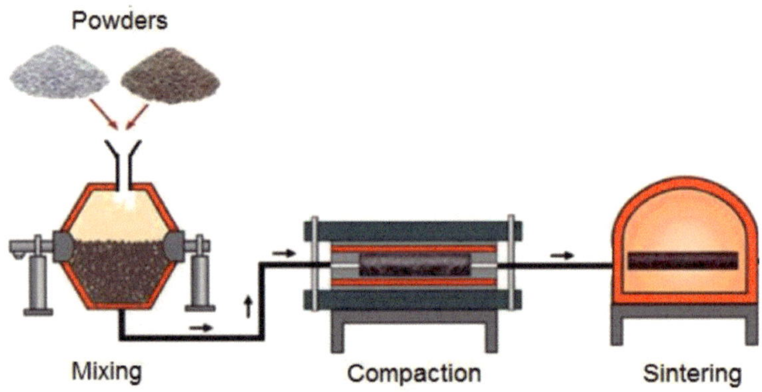

Figure 7.
Powder metallurgy process [7].

particles to bond to each other with minor change to the particle shape, which also allows porosity formation in the product when the temperature is well regulated. This method can produce high performance and low cost titanium alloy parts. The titanium alloy parts produced by powder metallurgy have several advantages such as comparable mechanical properties, near-net-shape, low cost, full dense material, minimal inner defect, nearly homogenous microstructure, good particle-to-particle bonding, and low internal stress compared with those titanium parts produced by other conventional processes [7].

4.2 Self-propagating high temperature synthesis

Self-propagating high temperature synthesis (SHS), shown in **Figure 8**, is another PM based process used to produce titanium alloys. The steps in this process include: mixing of reagents, cold compaction, and finally ignition to initiate a spontaneous self-sustaining exothermic reaction to create the titanium alloy [7].

Although the above PM processes are mature technologies for fabrication of bone implants they have difficulties of fabricating porous coatings on surfaces that are delicate or with complex geometries. In addition, these processes tend to produce brittle products because of cracks and oxides formed inside the materials. Further, the high costs and poor workability associated with these PM processes restrict their application in commercial production of bone implants. Consequently, new methods, based on additive manufacturing principles were developed [7].

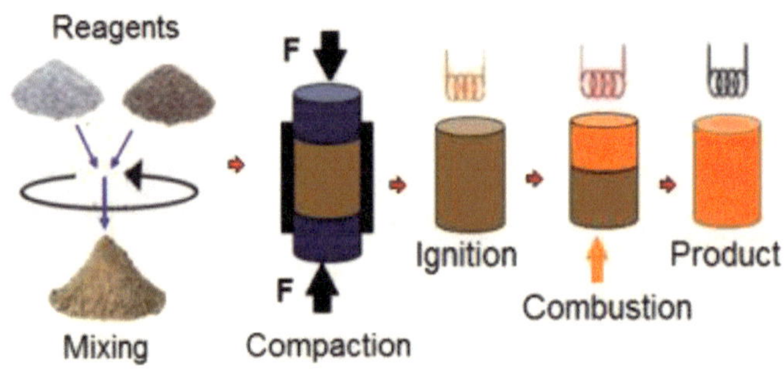

Figure 8.
SHS process [7].

5. Advanced methods for production of titanium alloys

The definitions of advanced methods of production is the use of technological method to improve the quality of the products and/or processes, with the relevant technology being described as "advanced," "innovative," or "cutting edge." These technologies evolved from conventional processes some of which have been developed to achieve various components of titanium base alloys and aluminides. Atomisation processes are among the most widely used cutting edge methods for production of titanium alloys [5].

5.1 Atomisation

Atomisation processes are used to make alloyed titanium powders. In these processes, the feedstock material is generally titanium, and the alloy powders produced are further processed typically to manufacture components using processes such as hot isostatic pressing (hip). As mentioned previously, it is generally believed that alloyed powders are not suitable for cold compaction using conventional uniaxial die pressing methods. Moreover, the inherent strength of the alloyed powders is too high, making it difficult to deform the particles in order to achieve desired green density. The atomisation processes produce relatively spherically shaped titanium alloy powders that are most suitable for additive manufacturing using techniques such as selective laser melting or electron beam melting. These spherical powders are also required for manufacturing titanium components using metal injection molding techniques. Typically, additive manufacturing and metal injection molding processes require particle sizes of powders to be in the range of 100 μm to ensure good flowability of the powder during operations. However, the challenge of the atomisation processes usually is that powders produced tend to have a wide particle size distribution, from a few to hundreds of micrometers. Examples of atomisation processes are gas atomisation and plasma atomisation processes described below [5].

5.1.1 Gas atomisation process

In the gas atomisation process, shown in **Figure 9**, the metal is usually melted using gas and the molten metal is atomised using an inert gas jets. The resultant fine metal droplets are then cooled down during their fall in the atomisation tower. The metal powders obtained by gas-atomization offer a perfectly spherical shape

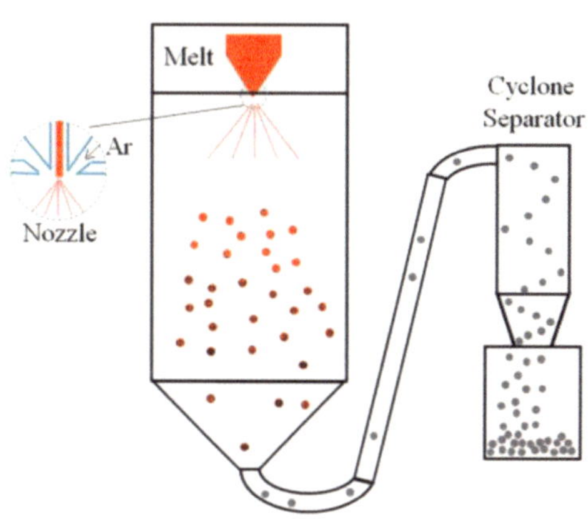

Figure 9.
Schematic diagrams of gas atomisation process [5].

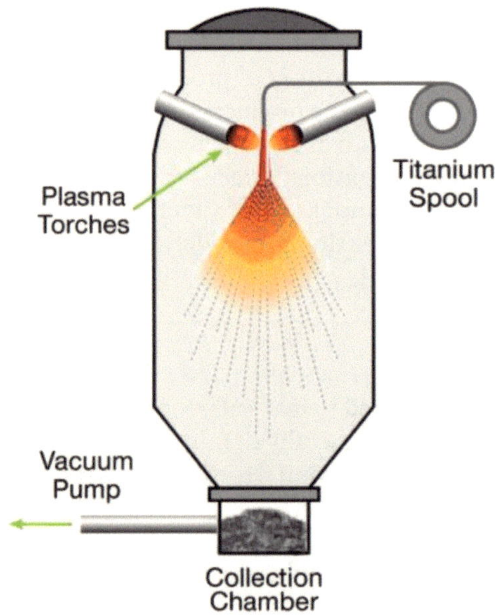

Figure 10.
Schematic diagrams of plasma atomisation process [5].

combined with a high cleanliness level. However, even though gas atomisation is, generally, a mature technology, its application need to be widened after addressing a few issues worth noting such as considerable interactions between droplets while they cool during flight in the cooling chamber, causing the formation of satellite particles. Also, due to the erosion of atomising nozzle by the liquid metal, the possibility for contamination by ceramic particles is high. Usually, there may also be argon gas entrapment in the powder that creates unwanted voids [5].

5.1.2 Plasma atomisation process

Plasma atomisation, shown in **Figure 10**, uses a titanium wire alloy as the feed material which is a significant cost contributing factor. The titanium alloy wire, fed via a spool, is melted in a plasma torch, and a high velocity plasma flow breaks up the liquid into droplets which cool rapidly, with a typical cooling rate in the range of 100–1000°C/s. Plasma atomisation produces powders with particle sizes ranging from 25 to 250 μm. In general, the yield of particles under 45 μm using the plasma wire atomisation technique is significantly higher than that of conventional gas atomisation processes [5].

6. Future methods for production of titanium alloys

The future methods for production of titanium alloys depend on the demand of these products and to what extend nature will be able to provide them. The demand for titanium alloys shall also influence the number and type of technological breakthroughs, the extent of automation, robotics' application, the number of discoveries for new titanium alloys, their methods of manufacturing, and new areas of application. Automation is an important aspect of the industry's future and already a large percentage of the manufacturing processes are fully automated. In addition, automation enables a high level of accuracy and productivity beyond human ability—even in hazardous environments. And while automation eliminates some of the most tedious manufacturing jobs, it is also creating new jobs for a re-trained workforce. The new generation of robotics is not only much easier to program, but also easier to use due to extra capabilities such as voice and image recognition during operations, they are capable of doing precisely what you ask them to do. The discovery of new titanium alloys, or innovative uses of existing ones, is essential for making progress in many of the technological challenges we face. This discovery can result in new synthesis methods of new alloy compounds and design of super alloys, theoretical modeling and even the computational prediction of titanium alloys. This discovery requires that new methods of manufacturing are developed. In light of this, "additive manufacturing" is being developed and this is viewed as a groundbreaking development in manufacturing advancement that offers manufacturers powerful solutions for making any number of products cost-effectively and with little waste. Examples of additive manufacturing technologies are cold spray, 3-D printing, electron beam melting, and selective laser melting. To fabricate alloy surfaces using these technologies, alloying elements are mixed thoroughly in the feedstock powder and the fabrication processes proceed as described in the following paragraphs [7, 8].

6.1 Cold spray

Cold spray (CS) process, schematically shown in **Figures 11** and **12** can deposit metals or metal alloys or composite powders on a metallic or dielectric substrate using a high velocity (300–1200 m/s) jet of small (5–50 μm) particles injected in a stream of preheated and compressed gas passing through a specially designed nozzle. The main components of a generic CS system include the source of compressed gas, gas heater, powder feeder, spray nozzle assembly, and sensors for gas pressure and temperature. The source of compressed gas acquires the gas from an external reservoir, compresses it to desired pressure and delivers it into the gas heater. Then, the gas heater preheats the compressed gas in order to increase its enthalpy energy. The preheated gas is delivered into the spray nozzle assembly whose convergent/divergent geometry not only converts the enthalpy energy of the gas into kinetic energy but also mixes the metal powders with the gas proportionately. The powder feeder meters and injects the powder in the spray nozzle assembly. The sensors for the gas pressure and temperature are responsible for regulating the preset pressure and temperature of the gas stream. The powder injection point in the spray nozzle assembly, the gas pressure, and gas temperature distinguish the low pressure-CS system (LP-CS) from the high pressure CS (HP-CS). In the LP-CS

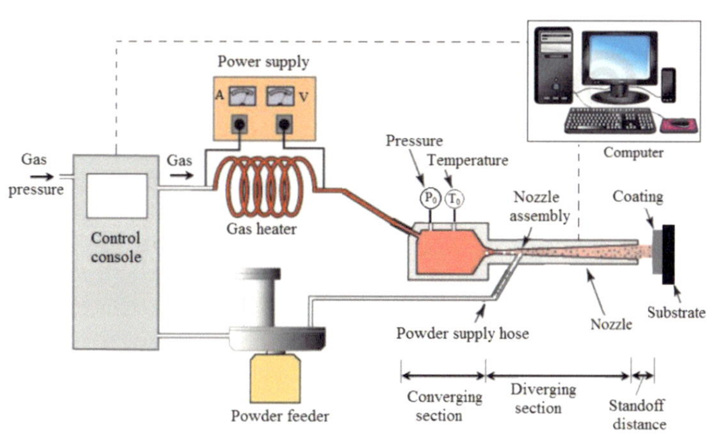

Figure 11.
Low pressure CS process configuration [8].

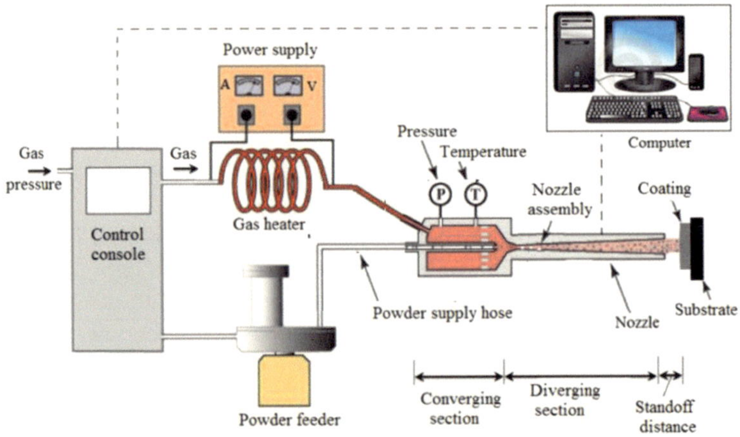

Figure 12.
High pressure CS process configuration [8].

Parameters	HP-CS	LP-CS
Working gas	nitrogen, air, helium	nitrogen, air, helium
Gas pressure, Mpa	0.5 – 5.0	0.5 – 1.7
Gas temperature, °C	20 – 1000	20 – 600
Gas flow rate, m^3/h	2 – 200	15 – 78
Maximum gas match no.	1 – 3	1 – 3
Powder flow rate, g/s	0.1 – 2.0	0.1 – 1.0
Particle size, μm	5 – 100	5 – 50

Table 1.
Operation parameters for CS systems [8].

system, the feedstock powder is injected in the downstream side of the convergent section of the nozzle assembly, while in the HP-CS system; the powder is injected in the upstream side of the convergent/diverging section of the nozzle assembly as illustrated in **Figures 11** and **12**. Several other parameters which contribute towards the distinguishing of the CS systems are summarized in **Table 1** [8].

6.2 3-D printing

3-D printing is an additive manufacturing method that applies the principle of adding material to create structures using computer aided design (CAD), part modeling, and layer-by-layer deposition of feedstock material. This cutting-edge technology is also called stereolithography, and is illustrated in **Figure 13** [8].

In this technology, the pattern is transferred from a digital 3D model, stored in the CAD file, to the object using a laser beam scanned through a reactive liquid polymer which hardened to create a thin layer of the solid. In this manner, the structure is fabricated on the desired surface. This method was proved in the laboratory setup is still being integrated in commercial set-up because 3-D printing is the most widely recognized version of additive manufacturing. For this reason, the inventors and engineers for this process have for years used machines costing anywhere from a few thousand dollars to hundreds of thousands for rapid prototyping of new products. It can be noted that all of the additive-manufacturing processes follow

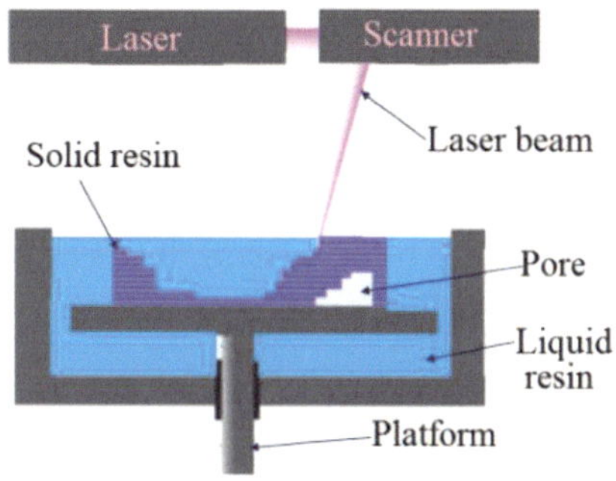

Figure 13.
3D-printing process [8].

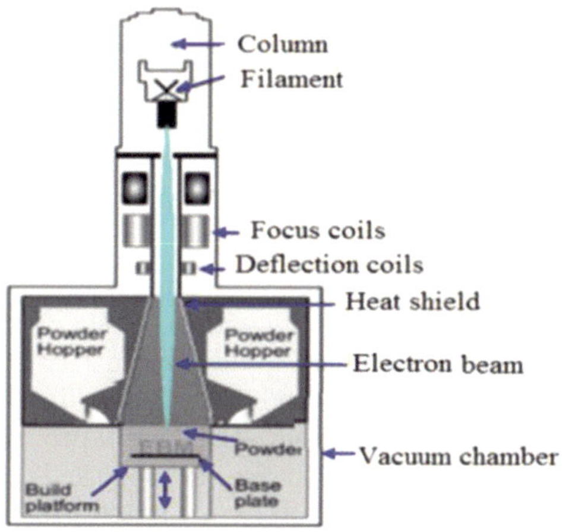

Figure 14.
Electron beam melting method [1].

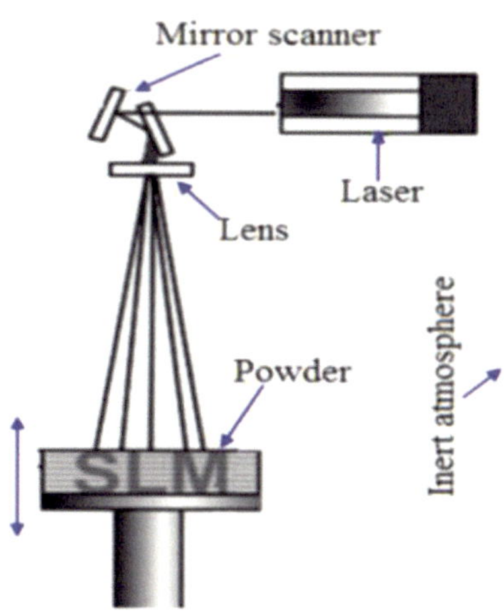

Figure 15.
Selective laser melting method [1].

this same basic layer-by-layer deposition principle but with slightly different ways such as using powdered or liquid polymers, metals, metal-alloys or other materials to produce a desired product [8].

6.3 Electron beam melting

Electron beam melting (EBM), shown in **Figure 14**, is one of the additive manufacturing processes which fabricated titanium coatings by melting and deposition of metal powders, layer-by-layer, using a magnetically directed electron beam. Though this method was proved to be successful, it has high set-up costs due to the requirement of high vacuum atmosphere [7].

6.4 Selective laser melting

Selective laser melting (SLM), shown in **Figure 15** is the second additive manufacturing method for titanium alloy coatings which completely melt the powder using a high-power laser beam. Similarly, this method is costly because it requires advanced high rate cooling systems. Moreover, the fluctuations of temperatures during processing negatively affect the quality of the products [1].

7. Conclusion

This chapter described the titanium as a metal that exists naturally with two crystalline forms. The chapter highlighted the properties of titanium metal that influence its application. The fact that titanium has advantageously unique properties that can be improved by alloying with other elements makes it to be preferred engineering material for future application in such areas as biomedical implants, aerospace, marine structures, and many others. The chapter discussed the traditional, current and future methods necessary to produce structures using titanium and titanium alloys. Further, the chapter suggested "additive manufacturing methods" as advanced methods for future manufacturing because they offer powerful solutions for making any type and number of products cost-effectively and with little waste. The examples of these methods are cold spray, 3-D printing, electron beam melting, and selective laser melting. Finally, the various processes used during fabrication of alloys using these methods were also presented.

Author details

Hamweendo Agripa and Ionel Botef*

1 Evelyn Hone College, Lusaka, Zambia

2 School of Mechanical Industrial and Aeronautical Engineering, University of the Witwatersrand, Johannesburg, South Africa

*Address all correspondence to: ionel.botef@wits.ac.za

References

[1] Sidambe AT. Biocompatibility of advanced manufactured titanium implants a review. Materials. 2014;7(12):8168-8188

[2] Yang R, Hao Y, Li S. Development and application of low-modulus biomedical titanium alloy Ti2448. Biomedical Engineering Trends. 2011;**10**:225-247

[3] Elahinia MH, Hashemi M, Tabesh M, Bhaduri SB. Manufacturing and processing of NiTi implants: A review. Progress in Materials Science. 2012;**57**:911-946

[4] Kopeliovich D. Titanium Alpha-Beta Alloys; 2018

[5] Fang ZZ, Paramore JD, Sun P, Chandran KSR, Zhang Y, Xia Y, et al. Powder metallurgy of titanium—Past, present, and future. International Materials Reviews. ISSN: 0950-6608 (Print) 1743-2804 (Online). Available from: http://www.tandfonline.com/loi/yimr20

[6] Ovchinnikov A, Smolyak Y, Dzhugan A, Yunusov E, Ianko T, Panov S. Technology of New Generation Titanium Alloys Powder for Additive Technology; 2018. pp. 101-107

[7] Hamweendo A, Malama T, Botef I. Titanium-nickel alloys for bone tissue engineering application via cold spray. In: International Conference on Competitive Manufacturing COMA16. Vol. 2016. Stellenbosch, South Africa. pp. 273-279

[8] Hamweendo A. Fabrication of porous structures using cold gas dynamic spray technology [thesis]. Johannesburg, South Africa: University of the Witwatersrand; 2017. pp. 1-198

Chapter 3

Microstructure and Mechanical Properties of Laser and Mechanically Formed Commercially Pure Grade 2 Titanium Plates

Kadephi Vuyolwethu Mjali and Annelize Botes

Abstract

The microstructure and mechanical properties of laser and mechanically formed commercially pure grade 2 titanium plates are discussed in this chapter. The microstructure of the as received parent material is compared to that resulting from laser and mechanical forming processes. Residual stress results from the two forming processes are analysed and bring to light changes brought about by these processes to the titanium used. The effect of the two forming processes on the mechanical properties is discussed, and the effect of process parameters on these properties is also argued in detail.

Keywords: laser forming, mechanical forming, residual stress, tensile testing, hardness testing

1. Introduction

The processing of engineering materials has become a specialist field, and this industry will continue to grow due to rising costs in raw materials which have forced many automotive and aviation industry suppliers to invest heavily in this field. In order to be relevant and competitive in today's industrial world, companies around the world are now forced to dedicate billions of dollars in profits to research and development. Many research centres are looking at titanium as a solution to some of the engineering challenges facing both automotive and aviation industries. Titanium is now finding favour with companies in pursuit of savings in fuel consumption and related improvements to mechanical properties. Savings in fuel consumption is achieved by reducing weight on aircraft and automobiles yet still meeting acceptable industrial norms and standards like improved structural integrity on the finished product. Improvements in engine and turbine design have also helped in the pursuit of fuel efficiency in these industries. In-depth research into the behaviour of titanium alloys under varying loading conditions is therefore essential in the quest to find more industrial applications of this metal. The last century saw a major development in processing and fabricating techniques. These developments were largely in part as a result of great emphasis placed in research and a continued search for improved methods in metal forming. This contributed to the development in forming techniques, materials, processing and understanding of changes a metal goes through during forming. There has always been room for improvement in the forming of materials due to the widespread use of forming operations in the automotive, aviation and shipbuilding industries. An in-depth study into the effects of laser and mechanical forming processes on the mechanical properties of commercially pure grade 2 titanium plates was conducted. This was achieved by producing a radius of curvature of approximately 120 mm on the plates with the aid of the mechanical forming machine. The plate samples were then subjected to mechanical testing to evaluate changes in mechanical properties. A Nd:YAG laser was used to replicate what had been achieved using the mechanical forming machine to bend titanium to the same radius of curvature. It was anticipated that this would lead to an extension of applications of laser forming and the possibility of increasing strength of thin commercially pure (CP) grade 2 titanium plates due to the heat treatment characteristics induced by the process. The laser forming study used established parameter settings which greatly influence the microstructure and bend radii. The intention of the study was to use both mechanical and laser forming to bend titanium plates to a final radius of curvature of 120 mm.

2. Commercially pure grade 2 titanium

Table 1 shows the chemical composition of the titanium used in this study which is weldable and formable and has excellent corrosion resistance properties. The tensile and yield strength values go up with grade number for pure grades.

Titanium can be cold rolled at room temperature to above 90% reduction in thickness without serious cracking [1]. Titanium undergoes allotropic transformation from the hexagonal close-packed (hcp) alpha phase to the body-centred cubic (bcc) beta phase at a temperature of 883°C. At room temperature, its properties are controlled by chemical composition and grain size. The presence of these elements determines the nature of the alloy and its chemical properties. The density of alpha titanium alloy falls between that of aluminium alloys ($2.7\,g/cm^3$) and steel ($7.8\,g/cm^3$) at $4.51\,g/cm^3$ as indicated. Due to the high-yield stress values of titanium, which are similar to steels and twice the strength of aluminium, makes this metal a choice in areas where weight is an important consideration [2]. An inhibiting factor especially in the automotive industry is the cost involved in using titanium as the main structural metal, whereas in the aviation industry, the manufacturers are able to include the cost of titanium in the final price of their products. The physical properties of CP titanium and properties like linear expansion coefficient, thermal conductivity and specific heat capacity playing a major role in the laser forming process are shown in **Table 2**.

2.1 Laser forming

Laser forming (LF) evolved from more mature, but less sophisticated thermo-mechanical forming processes. Specifically, manual application of an oxyacetylene

Grade	C	O_2	N_2 Max	Fe Max	H Max	Ti
Commercially pure titanium (as supplied)	0.005	0.155	0.009	0.04	0.003	Bal

Table 1.
Chemical composition of commercially pure grade 2 titanium in % wt.

Property	Linear expansion coefficient	Thermal conductivity	Specific heat capacity	Electrical resistivity	Alpha/beta transform temperature	Young's modulus	Shear modulus	Poisson's ratio	Density
Alpha titanium	8.36×10^{-6} K^{-1}	14.99 W/m.K	523 J/kg.K	5.6×10^{-7} Ohm.m	882.5°C	115 GPa	44 GPa	0.33	4.51 g/cm^3

Table 2.
Physical properties of pure titanium.

torch for forming steel plates for the ship building industry has been used for some time and is seen as the precursor to LF [3].

The laser forming process has become a choice to fabricators of metallic components and as a means of rapid prototyping and of adjusting and aligning [4]. Laser forming is of importance to industries that previously relied on expensive stamping dies and presses for prototype evaluations. Industry sectors making use of this process include aerospace, automotive, shipbuilding and those in microelectronics. The laser forming process involves no mechanical contact, which is a requisite in mechanical forming and is considered a virtual manufacturing kind of method. The laser forming process can be used to produce predetermined shapes. The process results in minimal distortion on the formed components [5]. The laser forming process can produce metallic, predetermined shapes with minimal unwanted distortion, and investigations are also ongoing in the removal of unwanted distortion resulting from the procedure.

A successful and significant research in the laser forming of materials needs a good understanding of thermal transfer concepts as they play a crucial role in the process. Concepts like conduction and thermal radiation need to be understood fully to balance all the process variables. Thermal radiation is the transfer of energy by electromagnetic waves, whereas thermal conductivity, on the other hand, is the property a material possesses indicating its ability to conduct heat. Thermal conductivity of titanium is lower than most competing metals like steel, magnesium and aluminium. This means that in order to cause changes in the microstructure, a higher intensity of heat would have to be emitted by the heat source and in this scenario by the laser. The ability of the plate material to absorb and transfer heat is the major underlying factor. This factor plays a major role in the forming of plates as the effect of conduction affects the microstructure, thereby influencing the mechanical properties. The heat flux (power density), which plays a considerable role in the laser forming process, is the amount of energy flowing through a particular surface area per unit of time and is represented by the following formula:

$$q = \frac{Q}{\pi r^2} \tag{1}$$

where q is the heat flux, Q is the laser beam power (W), r is the beam radius (m), and π is the constant.

According to Ion et al., a large number of variables influence the interaction between a laser beam and a material; over 140 variables can be identified for welding alone. In this instance, the power density will be considered when a beam is switched on. The heat flow becomes steady state, and the energy absorbed by the surface is balanced by that conducted heat into the plate, and the temperature field becomes constant. The principal process variables are the beam power, the beam radius and material properties. The power density can be increased four times by quadrupling the power or by reducing the beam radius to a half. When this variable group is identified, a smaller subset of experiments can be undertaken to establish that the power density determines the peak surface temperature attained. These factors determine the principal mechanism of thermal interaction—which could either be heating, melting or vaporisation [6]. The laser powers used, the thermal conductivity, the line energy, scanning velocities, beam interaction time and also the heat flux (power density) generated during the laser forming process are all shown in **Table 3**.

Power and scanning velocities were adjusted during the preparation of plate specimens used in this study and the other given parameters resulted from these adjustments (beam interaction time, heat flux and thermal gradient). The heat flux formula was considered for analysis in order to understand the concepts involved in this process. The laser power ranged from 1.5 to 3.5 kW for the specimens evaluated, and an increase in power resulted in an increase to the heat flux and line energy generated. With the arrangement used, the samples are not clamped in any way, and the line heating application alternates in succession from each end incrementally moving towards the centre of the plate. The open mould method shown in **Figure 1** was used in the laser forming of the CP grade 2 titanium plates.

The beam interaction time was an important factor in the analysis of the resulting microstructure and can be determined by the formula

$$t = \frac{2r_b}{v} \tag{2}$$

The variables $2r_b$ and v represent the beam radius and the scanning velocity, respectively. The power density, beam radius and beam interaction time play a considerable role as they determine whether the material will be cut, welded, melted or hardened. The heat flow in laser processing can be complex, but for many processes it may be approximated to three fundamental conditions: steady state, transient or quasi-steady state. Fourier's first law describes steady state conditions as

$$F = -\lambda \Delta T \tag{3}$$

where F is the heat flux ($W/m2$), ΔT is the thermal gradient (K/m), and λ is the thermal conductivity ($W/m.K$). In this state, the temperature field does not change with time at a location in a material. The thermal gradient is a physical quantity that describes in which direction and at what rate the temperature changes most rapidly around a particular location. The thermal gradient can lead to different amounts of contraction in different areas, and if residual tensile stresses become high enough, flaws may propagate and cause failure. A lower thermal gradient may cause bending in other engineering materials but due to differing thermal conductivities may not work in other materials. This means that each engineering material needs to be isolated in the analysis of its physical properties. For example, what may work for steel may not be applicable to titanium due to different thermal conductivities of the two materials. Line energy is a concept used by engineers and scientists in laser forming to control bending characteristics of plates. According to Magee, the energy input to the sheet-metal surface critically affects the nature of the process and forming mechanisms which take place [7].

The line energy specified by Magee is a function of laser power and the scanning velocity. In determining the process parameters for the experimental exercise, four sets of power levels believed to result in the desired curvature were chosen and are listed in **Table 3** and discussed here. The laser forming process produces large thermal gradients that could either bend or shorten the material. The bending or shortening of the material is a result of the line energy produced by the laser and is given by the formula

Laser power (kW)	Thermal conductivity (W/m.K)	Line energy (kJ/m)	Scanning velocity (m/min)	Beam interaction time (sec)	Heat flux ($\times 10^6$ W/m^2)	Thermal gradient ($\times 10^3$ K/m)	Beam diameter (mm)	Average radius of curvature (mm)
1.5	15	35	2.62	0.0091	13.3	221	12	180.1
1.5	15	47	1.9	0.0122	13.3	221	12	150.3
2.5	15	90	1.7	0.0141	22.11	1474	12	134.3
3	15	90	2	0.0121	26.53	1769	12	118.4
3.5	15	90	2.3	0.0101	31	2064	12	106.1

Table 3.
The various parameters involved in the laser forming process.

Figure 1.
Laser formed CP grade 2 titanium with the open mould arrangement.

$$L = \frac{P}{v} \tag{4}$$

where P represents laser power in Watts (W) and v represents the scanning velocity in metres per minute, respectively. Line energy is the most important variable in the laser forming process. It was also decided to determine what influence the variation in power levels would have on the microstructure and the mechanical properties of titanium [8]. There is a widespread belief that a line energy threshold must be exceeded in order to commence with permanent deformation by the temperature gradient method (TGM) [9]. A line energy of 90 kJ/m was considered after unsuccessful attempts to bend samples at a power of 1.5 kW where the line energies of 35 and 47 kJ/m, respectively, were used. For powers ranging from 2.5 to 3.5 kW, the line energy was kept constant at 90 kJ/m, and the scanning velocities were adjusted to suit the required line energy, in this instance 90 kJ/m.

The prime pocket monitor shown in **Figure 2** was used in this study to measure the laser power projected on the surface of titanium specimens. Readings were taken to fully understand the incident power hitting the titanium plate surface. As an example for a laser power setting of 3500 W, the pocket monitor reader would show 3250 W. This value indicates a 10% loss in power on the irradiated sample [8]. This assisted in understanding and acknowledging the presence of losses in laser irradiation in material processing. For the purpose of this study, the losses were ignored and not taken into consideration in the analysis.

2.2 Mechanical forming

Metals are used extensively as engineering materials in part because of their ability to deform plastically. Various forming processes are used to form

Figure 2.
Prime pocket monitor.

engineering materials to desired shapes and sizes. These forming operations generally occur after the metal is cast, so it is important to understand how forming operations interact with pre-existing casting defects. Most metal forming operations reduce the severity of casting defects, such as microporosity, and break up coarse particles, such as non-metallic inclusions, that form during solidification. The mating die method was used to bend titanium alloy plates to the desired radius of curvature. The mating die method of stretch-draw forming involves an upper and lower die block mounted in a hydraulic press bed. The workpiece is securely held in tension by movable grippers. Yield stress of the finished part may be increased as much as 10% by the stretching and cold working operations. Shown in **Figure 3** is how the bending of titanium plates was achieved using the mechanical forming machine and also the resulting shape.

The objective of this study was to bend a flat plate of titanium to a radius of 120 mm, and this would help in understanding the principles behind the mechanical forming process. The study also aimed at comparing mechanical to laser forming with regard to the microstructure and mechanical properties of the material and any changes that happen thereafter as a result of both forming operations.

2.3 Tensile test

The mechanical properties of CP grade 2 titanium alloy vary with its grade as indicated in **Table 4**. CP grade 2 titanium plate specimens were evaluated according to the American Society for Testing and Materials (ASTM) E8/E8M test method. The tensile test was performed on the parent material (CP grade 2 titanium plates) in accordance with ASTM E8, using the Hounsfield machine.

The resulting data were made available using computer software of the machine. The table shows average values taken from both the transverse and longitudinal directions of the plate. The ultimate tensile strength (the maximum engineering stress in tension that may be sustained without fracture) is given as 452 MPa, the yield 338 MPa and a percentage elongation of 28%.

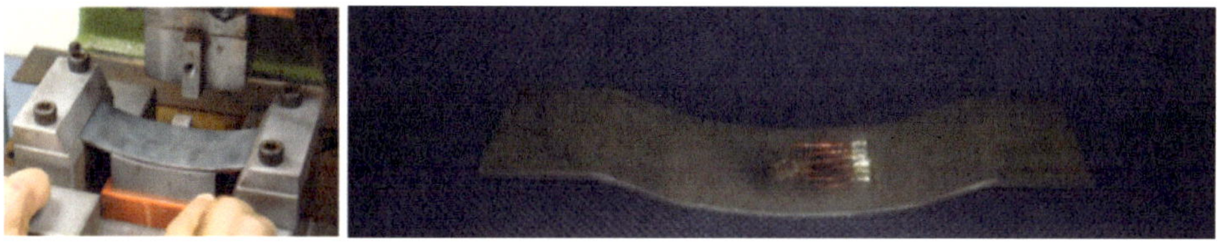

Figure 3.
Mechanical forming.

Alloy	E (GPa)	$\delta_{0.2}$ (MPa)	UTS (MPa)	Elongation (%)
Grade 1	105	170	240	24
Grade 2	105	275	345	20
Grade 2 (current study)	**105**	**338**	**452**	**28**
Grade 3	105	380	445	18
Grade 4	105	480	550	15

Table 4.
Mechanical properties of CP grade 2 titanium alloy.

2.4 Microstructure

Figure 4 shows microstructures from the parent material specimens, and this material has equiaxed α-grains usually developed by annealing cold-worked alloy above recrystallization temperature. The microstructure has shown results from the manufacturing process of CP grade 2 titanium, which cannot be altered in the plates without the addition of heat or cold deformation processes.

The microstructure of mechanically formed plates contains the same equiaxed alpha grains found in the parent material. Mechanical forming produces no heat, and therefore the similarities in microstructure are to be expected. There were no major changes to the microstructure as a result of this process when compared to the as received material [10]. The microstructure of a mechanically formed plate is shown in **Figure 5**.

Figure 6 shows the fine structure of titanium from the plates irradiated at a power of 1500 W using line energies of 35 and 47 kJ/m, respectively. There is a variation in the depth of the heat-affected zone (HAZ) for both line energies [10]. The unaffected material in both cases has equiaxed α-grains similar to those in the parent material. Based on microstructural observations, it becomes clear that the temperature generated at **35 kJ/m** was less than that generated at **47 kJ/m** as it could

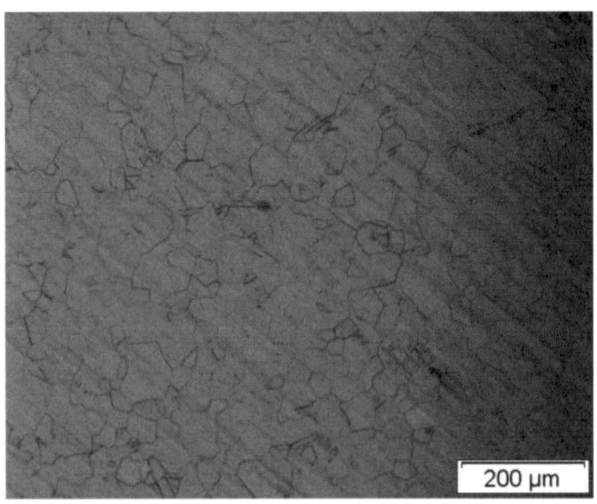

Figure 4.
As received material [parent material].

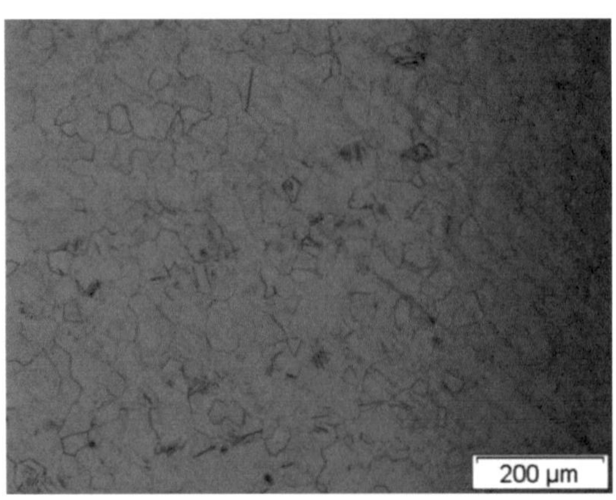

Figure 5.
Mechanically formed microstructure.

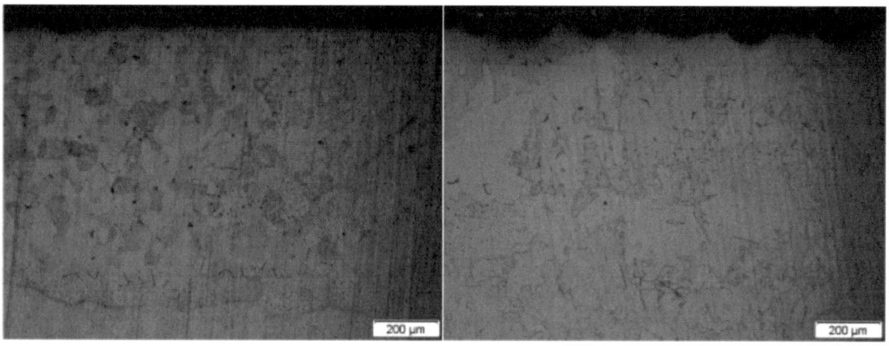

Figure 6.
Microstructure of laser formed plates [1.5 kW, 35 and 47 kJ/m].

not penetrate as deep to effect changes deeper in the microstructure away from the laser-irradiated surface.

The resulting microstructure from a line energy of 35 kJ/m points to a higher scanning velocity. The thermal energy from the laser managed to effect changes to a quarter of the plate's thickness [10]. The cycle took around 18 minutes to irradiate all the 10 plates in each batch at this power and line energy setting. The laser forming process resulted in a semi-circular-shaped heat-affected zone in the lower line energies (**35** and **47 kJ/m**). The area not affected by the heat in both cases shows smaller grains than those in the heat-affected zone. Grain size depends largely on temperature attained during the laser forming process, and grain growth proceeds more quickly as temperature increases. All the figures shown clearly reveal the influence of temperature on the microstructure, and the portion affected by laser energy has grains which are much bigger than those not affected by heat [10]. This is the reason why there is variation in the microstructures.

Figure 7 shows a major change in the microstructure of CP grade 2 titanium plates with enlarged primary α-grains and enlarged β-grains (**2.5 kW**). The structure consists of much bigger equiaxed alpha grains in the structure. The resulting microstructure is a result of thermal energy developed by the process parameters on the plates irradiated. Thermal energy is the main initiator in microstructural layout in all the laser formed plates. The microstructure managed to change only halfway through the plate which explains why there was minimal bending on the plates irradiated. Alpha titanium is cooling rate sensitive as seen by differences between the top section (laser-facing side) and the middle section. The microstructure

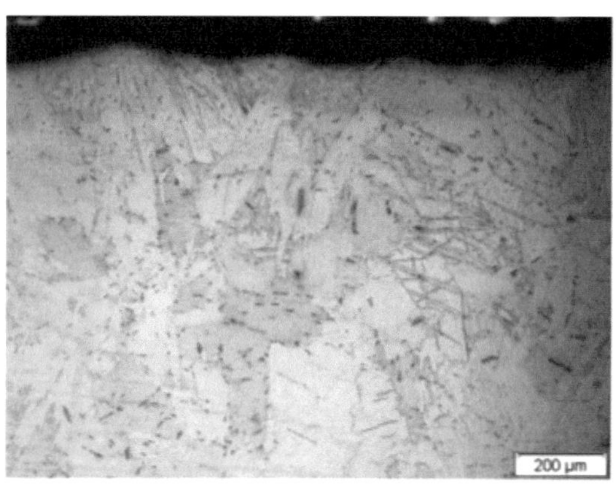

Figure 7.
Microstructure of a laser formed plate [2.5 kW, 90 kJ/m] [10].

formed as a result of the heat and cooling rate is not the same throughout the sample as witnessed on the as supplied parent plate. There are differences in microstructure between the top, the middle and bottom sections of the plate samples.

On the section of the plate closest to the source of laser irradiation and as thickness of the plate increases, the effect of thermal energy diminishes. The different microstructures shown are also an indication of different hardness values. The forming parameters at this power level led to plastic deformation on the laser-facing side. Before getting to plastic deformation, the grains were similar to those of the as received material (parent and mechanically formed plates). The scanning velocity used here happens to be the lowest in this study. The low scanning speed meant that the laser got more time to effect changes per unit area of the material resulting in the microstructure shown. The cooling of the plates also contributed to the microstructure. All the plates were naturally cooled. Thermal measurements have also shown the effect of the scanning velocity on the material. In multiple scan scenarios, each scan effects change on the microstructure. Differences in microstructure are brought about by the laser intensity power of **2.5 kW** which makes a significant change in the microstructural layout [10].

Figure 8 also shows the microstructure of a sample irradiated at **3 kW**, and with this plate an increase in power results in gradual change to the microstructure of titanium. The microstructure has much bigger equiaxed-α (alpha) and (beta) β-grains compared to a power of **2.5 kW** and the supplied parent material. The initial microstructure has an effect on the mechanical properties of titanium. During the process, changes in temperature affect the microstructure which in turn influences the mechanical properties of titanium. The changes in temperature and cooling rates also play a role in resulting mechanical properties. The high temperatures attained effected the top and bottom sections of the plates. An increase in power from 2500 to 3000 W meant that scanning speeds had to be adjusted in order to get to a line energy of 90 kJ/m. This reduced the time taken to irradiate the batch of samples. The heat flux increases by about 18% when the power is adjusted to 3000 W. There was also a reduction of 18% to the process time. The changes in heat flux indicate higher temperatures on the plate surface [10]. The alpha and beta grains are bigger closer to the centre of the irradiated plates and elongated closer to the laser-facing surface. The thermal energy generated resulted in different microstructures between the top and bottom halves of the sample. It should also be eminent that with an increase in power from 2.5 to 3 kW, there is a reduction in time taken to achieve irradiating the plate samples. The altering of power from 2.5 to 3 kW results in an 18% increase in the amount of heat flux generated and a 19%

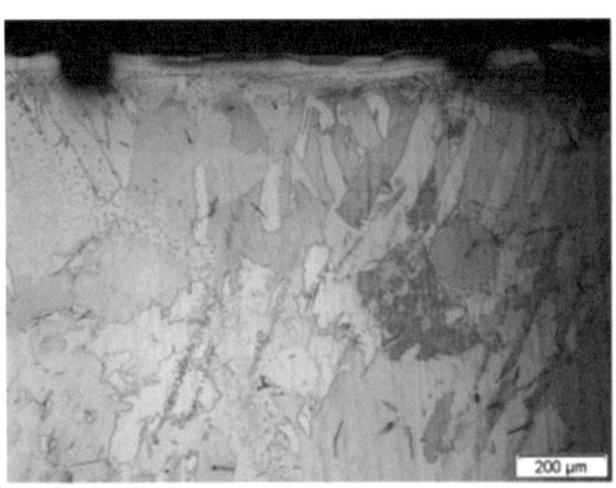

Figure 8.
Microstructure of a laser formed plate [3 kW, 90 kJ/m] [10].

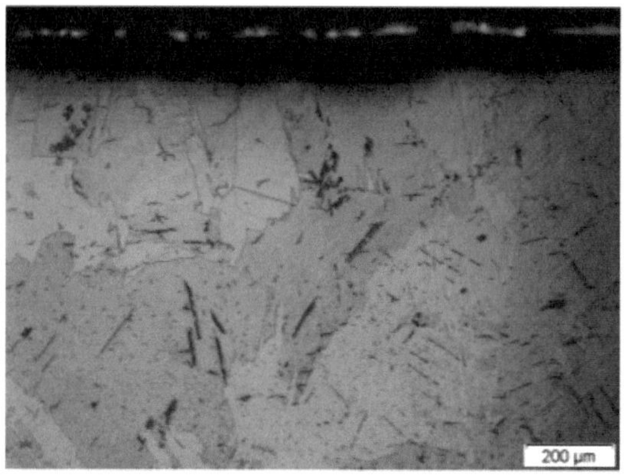

Figure 9.
Microstructure of a laser formed plate [3.5 kW, 90 kJ/m] taken from the top surface.

increase in scanning velocity. The change in scanning speed was done to achieve a line energy of **90 kJ/m** and resulted in the decline of process time by 18%. These numbers show that changes in the microstructure are to be expected as the new heat flux generated amounts to higher temperatures on the surface of the plate. In as much as cooling rates determine the resulting microstructure on titanium alloys, the processing temperatures also play an important role as well [10].

Figure 9 shows the microstructure layout as a result of laser forming at the highest power setting. The grains from this setting were the biggest in all the plates evaluated in this study. Twin bands can be seen throughout the microstructure of the plate. All sections of the plate had different grain sizes attesting to different cooling rates in the plate. The surface facing the laser did not cool down at the same time as the opposite side of the plate [10]. Acicular alpha can also be seen on the side opposite the laser-irradiated surface. The complete laser irradiation for these samples took about 20 minutes, which explains the changes in the microstructure when compared to a power of 2.5 and 3 kW, respectively. Changing the power from 2500 to 3000 W resulted in a 40% increase in heat flux. The heat flux increased by 16% from a power of 3000–3500 W. This means that by changing the power values, there was a related increase in the heat transferred per unit area, per unit time. These changes contributed to changes in the accompanying microstructure and mechanical properties. The study on the microstructure and mechanical properties helped in understanding the behaviour of titanium in different forming scenarios. The information gathered also made it easier to analyse the hardness results.

2.5 Hardness

The hardness number is a resistance for the local plastic deformation, and the hardness is closely related to residual stresses [9]. The average Vickers hardness obtained for the parent material is 160 $\pm$ 5Hv0.3, and whilst the average hardness number for the parent material is higher than that obtained in mechanically formed samples, the laser formed specimens show higher values. The average hardness results of the mechanically and laser formed CP grade 2 titanium specimens are shown in **Table 5**.

Mechanically formed plates did not behave like laser formed samples as there was a slight increase in hardness moving away from the top section resulting in an average hardness of 130 $\pm$ 5Hv0.3. This is a result of changes in the material structure caused by the die during mechanical forming. The microstructure of plates irradiated at **1.5 kW (35 kJ/m)** indicate that heat energy could only penetrate

Material	Parent (as supplied)	Mechanically formed	1.5 kW (35 kJ/m)	1.5 kW (47 kJ/m)	2.5 kW (90 kJ/m)	3 kW (90 kJ/m)	3.5 kW (90 kJ/m)
Average Vickers hardness	160	130	176	171	410	349	311

Table 5.
Hardness profile of laser and mechanically formed plate samples.

to a third of the depth of the sample (changing a small portion of the microstructure). Due to the low amount of heat generated, there was a minor change in the microstructure, and this translated to minimal changes in the hardness values. The increase in line energy from 35 to 47 kJ/m also contributed to an increase in hardness. The increase in hardness values could be traced back to the change in the size of the microstructure grains when compared with the parent material [10].

On examining the microstructure of specimen irradiated at **47 kJ/m** (1.5 kW), the change in the structure is more remarkable than the plate samples irradiated at **35 kJ/m** (1.5 kW). The processing speed at **47 kJ/m** was slower, making it possible for the laser to effect changes on the microstructure on a much improved scale resulting in a bigger heat-affected zone. Values obtained at this power level and line energy are higher than those obtained from a line energy of **35 kJ/m**. These values also show the importance of thermal energy in the laser forming process. The power of **2.5 kW** had the highest average hardness values in all the plate samples evaluated. This could be linked to the low scanning velocity at this power level. More heat was dissipated per unit area per unit time resulting in the high hardness values. Hardness values obtained at a power of 2500 W had high values than that of the parent material. Process parameters at this power level proved to be the optimum settings for this study. For those engineering applications in need of improved hardness properties on this grade of titanium, these settings could be used. The optimum settings resulted in the highest value of Vickers hardness in this study at 410Hv0.3. The hardness value obtained shows a 100% increase in hardness when compared to both the parent and mechanically formed plates [10].

The reduction in hardness values at this power setting could be traced back to the grain structure found in samples irradiated. The microstructure contained acicular alpha and beta phases which have a significant effect on the mechanical properties of titanium. An average Vickers hardness value of 349Hv0.3 was obtained at this power setting, and it was the lowest on the samples evaluated. Plates irradiated at 3000 W had the hardness value of 349Hv0.3 in plates formed at a line energy of 90 kJ/m. The same plates showed a marked improvement in hardness at the middle section of the plates. The forming process effected physical changes on the surface of the plates. These changes translated to changes in the hardness of the material. The results show an improvement of more than 100% when compared with the as received material. These changes also made the material hard to polish during the preparation of residual stress samples [10].

A hardness value of 311Hv0.3 was obtained at a power setting of 3500 W. This is the third highest value in samples irradiating a line energy of 90 kJ/m. The size of grains and their structure were different when compared to other laser formed plates. Readings taken from the top section of the laser-irradiated side indicate a considerable increase in the average hardness of titanium. An average Vickers hardness value of 311Hv0.3 was obtained from the top section which indicates a 40% increase in the hardness of titanium. The Vickers hardness readings taken closer to the surface show increased hardness values which are much higher than those obtained from the parent plate by a bigger margin. The improvement in hardness as a result of the laser forming process could help in the preparation of titanium for other engineering applications in need of hardened titanium plates [10].

2.6 Residual stress

The graphs plotted from the analysed plates were a result of residual stress information gathered by the MTS3000 machine on each plate sample evaluated. Comparisons are made between the plates based on the graphs obtained. The relieved strain from the parent material differs to that obtained from other evaluated plates. **Figure 10** shows relieved strain measured on the parent material, and all the micro-strain values (ε_1, ε_2, ε_3) show a slight reduction in strain as the depth of the hole increases.

The parent material shows minimum values in both residual stress and strain. Even when the drill depth increases, residual stress and strain remain constant. The graph obtained is totally different when compared to other plates evaluated in this study. With the other power levels in laser formed plates, there were changes in residual stress and strain with changes in drill depth [8]. This figure also shows an even distribution of residual strains on the material, and, unlike the laser formed plates, it seems possible that the temperature gradient on the parent plates during fabrication was not steep. The residual strains are not modified in any way but result from the manufacturing procedure used to produce titanium. The other forming operations witnessed in the study show a marked change to the residual stress/strain distribution. Residual stress from as received parent material shows steep residual stress versus drill depth gradient. The gradient is typical of stress induced by the manufacturing process. Surface residual stress is of high importance to mechanical design engineers as they show areas of high residual stress. The high residual stress areas help contribute to fatigue failure of the material [8]. All values obtained in the analysis of residual stress and strain of CP grade 2 titanium plates are shown in **Table 6**, and results obtained allude to the performance of these plates during fatigue testing.

The readings obtained from the parent material form the base for the analysis of residual stress, and strain results for the forming process utilised in this study. Results from the parent material show a difference between the maximum and minimum stresses of 12.9 MPa which is tensile. The stress values also give an indication as to why the parent material performed better than other plates during fatigue testing. The laser formed plates showed higher values of stress than both mechanically formed and the parent materials. The effect of these stresses is

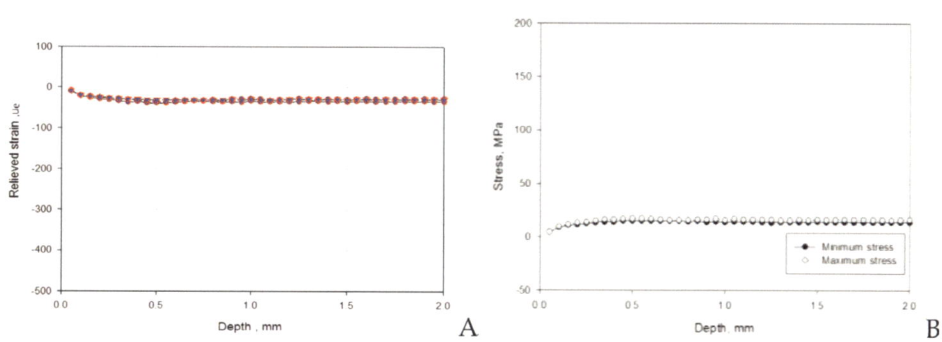

Figure 10.
Relieved strain (A) and stress (B) for parent material.

Samples	Minimum strain (με)			Maximum strain (με)			Minimum and maximum stress (MPa)	
	ε_1	ε_2	ε_3	ε_1	ε_2	ε_3	σ_1	σ_2
Parent material	−39,	−37	−32	−9.6	−9.81	−9	**4.1**	**17**
Mechanically formed	−111,	−59	−33	−16	−12	−4	**3.1**	**41.1**
1500 W (35 kJ/m)	−81	−180	−284	5	0.7	3	**−2.9**	**116**
1500 W (47 kJ/m)	−68	−245	−427	29	17	16	**−14**	**188.3**
2500 W	−199	−352	−443	−21	−27	−33	**11.4**	**181.8**
3000 W	−129	−276	−401	1.5	−5	−5.71	**−0.3**	**176.9**
3500 W	−166	−290	−403	−3	−2.61	−2.51	**1.4**	**181.9**

Table 6.
Residual stress and strain results [8].

therefore evident in fatigue testing and is documented in the results obtained [8]. The mechanical forming process resulted in minor changes to the relieved stress and strain, when compared to the parent material results. The mechanical forming process rearranges the residual stress and strain in the parent material. The term rearrange is applicable in this scenario as the material had residual stress within, prior to both forming processes. Some engineering applications encourage the presence of residual stresses within the material. The changes in residual stress are due to physical changes in the material as a result of laser and mechanical forming. Manufacturing processes introduce residual stress into mechanical parts, thereby influencing fatigue behaviour. The influence of all the forming operations is well documented in the analysis of fatigue results. The only difference between these processes is the intensity at which each forming process transpires. There are variations from process to process as witnessed in this study between mechanical and laser forming processes. After the attainment of maximum stress, there is a reduction in stress as the depth increases [8].

The mechanically formed plates had higher residual stress values than the parent material at 41 MPa. This is a 54% increase in stress when compared to the parent material. The difference in stress between maximum and minimum stresses was 38 MPa, a 66% improvement when compared to the parent material. These results had an influence on the fatigue results of the material. The graphs also show changes in residual stress with each forming process. There are similarities in residual stress between the parent material and the mechanically formed plates. The stress peaks at about 0.5 and 0.7 mm and then taper as maximum depth is approached. Based on results obtained from the parent material, forming moves the location of maximum and minimum principal stress closer to the surface. The low line energies had minimum effect on the residual stress distribution in the titanium plates [8] (**Figure 11**).

On the laser formed plates, there is a relative increase in the strain relaxation curve when compared to the parent and mechanically formed plates. In laser formed plates due to the physical changes in the material, there is a modification in the residual stress and strain due to phase transformation. The phase transformation is due to the intense heat from the laser and effects of the temperature gradient mechanism [8]. As witnessed on other laser formed plates, there is an increase in relieved strain as the line energy increases. The effect of deformation compatibility, as a result of internal stresses, is evident on the laser formed plates, and unlike mechanical forming, the effects of heat energy are evident on the tested specimens.

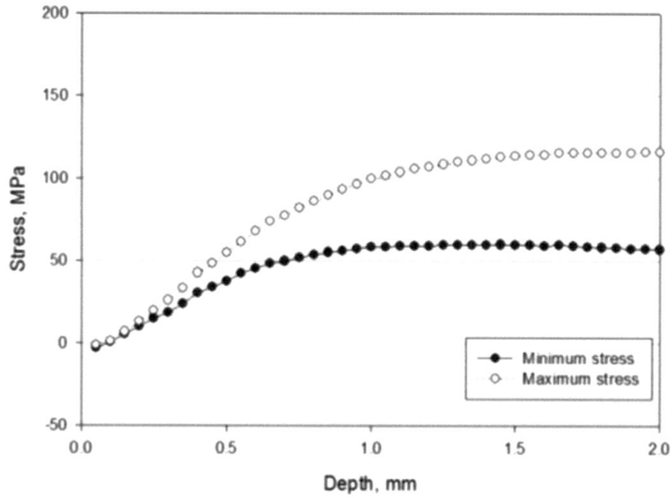

Figure 11.
Relieved stress, laser formed plates (1.5 kW, 35 kJ/m).

The laser forming process was carried out in such a way that there was an overlap on the scan tracks, meaning some portions of the laser-irradiated specimens did not get direct heat energy from the laser but were exposed to its effects. This resulted in large thermal gradients in the material contributing to an increased presence of internal stresses in the plates. The laser forming process has the ability to move the location of the maximum stress within the specimens as witnessed in all the laser formed specimens.

For the parent and mechanically formed specimens, the location of maximum principal stress was between 0.5 and 0.7 mm, respectively. With the laser formed plates, the location of maximum principal stress is between the depths of 1.5 and 2 mm. The changes in redistribution of residual stress are due to the thermo-mechanical properties of the laser forming process. For a power of 1500 W and a line energy of 35 kJ/m, the maximum stress attained was 116 MPa (T) and a minimum stress of 2.9 MPa(C). This maximum stress was the lowest in all laser formed plate samples. Maximum and minimum residual stress values do not decrease with changes in depth as witnessed with the parent plate. The changes in line energy change the location of maximum and minimum residual stress [8]**.** The line energy generated managed to penetrate and force a change on the microstructure of CP grade 2 titanium. Based on the microstructural analysis, there is a noticeable difference in microstructure between the line energies developed at a power of 1.5 kW (**35** and **47 kJ/m**).

The change in line energy from 35 to 47 kJ/m can be seen on the residual stress and strain results. With the line energy of 47 kJ/m, the relieved strain starts positive and ends negative due to a surge in gauge 2 and 3. These changes are due to the effects of laser forming which greatly influence the distribution of residual stress and strain. Changes in residual stress are also dependent on the process parameters and the line energy and heat flux generated. The line energy and heat flux are responsible for the phase transformation in the physical properties of the material. Titanium changes phase at a temperature of 883°C, and it appears that temperatures exceeding this value were reached during the laser forming process. The thermal gradient is the same as that obtained at an energy of 35 kJ/m. The same goes with values in heat flux which remain constant. The only difference is brought about by changes in scanning speed and beam interaction time. Changes in line energy caused variations in minimum and maximum residual stress values [8].

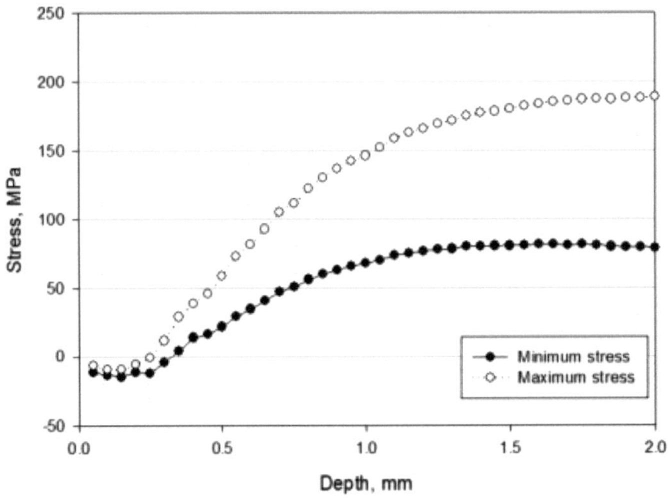

Figure 12.
Laser formed plates (1.5 kW, 47 kJ/m).

The maximum and minimum stress was the highest in all the specimens evaluated at 1.5 kW and is shown in **Figure 12** above. The increase in residual stress resulted in a reduction to fatigue life in laser formed specimens. There is a steady increase in both maximum and minimum principal residual stress and strain as line energy increase. These changes are influenced by changes in temperature which also affect the microstructure. Differences in residual stress are a result of different scanning speeds. The line energy of 47 kJ/m was obtained after adjusting the scanning velocity [from 2.6 to 1.9 m/min]. The change in speed meant there was an increase in beam interaction time causing more physical changes to the material. More time was therefore available per unit area per unit time to cause changes to the material. The power setting of 2500 W had a slower scanning velocity, a high heat flux, a higher line energy and minimal beam interaction time. These plates also experienced a higher thermal gradient which influenced changes in residual stress and strain [8].

Changes in microstructure also influenced the distribution of residual strains. The variations in thermal gradient between a power of 1500 and 3500 W caused major changes to the microstructure and led to a rise in non-uniform thermal strains, whose effect became hyperbolic when the material is elastically stiff and has a high-yield strength. The variations in temperature caused changes to the resulting mechanical properties. This means that the material properties are largely dependent on temperature. The higher the temperature, the greater will be the change in material properties [8].

The microstructure of the laser-irradiated specimens' changes as the depth of the specimen increases moving away from the laser-irradiated surface. The change in line energy to **90 kJ/m** resulted in an increase to the maximum principal stress which continued being in tension. Unlike the power of 1.5 kW, both maximum and minimum principal stresses start as being in tension and not compressive closer to the irradiated (laser-facing) side [8] (**Figure 13**).

On plates irradiated at 2500 W, the maximum and minimum residual stress was 182 MPa (T) and 11 MPa©, respectively. The difference in stress was 170 MPa, and the maximum stress is obtained at a depth of 1 mm. High residual stress had a negative effect during fatigue testing, as the material had deformed plastically. There is an alteration in the thermal gradient at this power, and the scanning

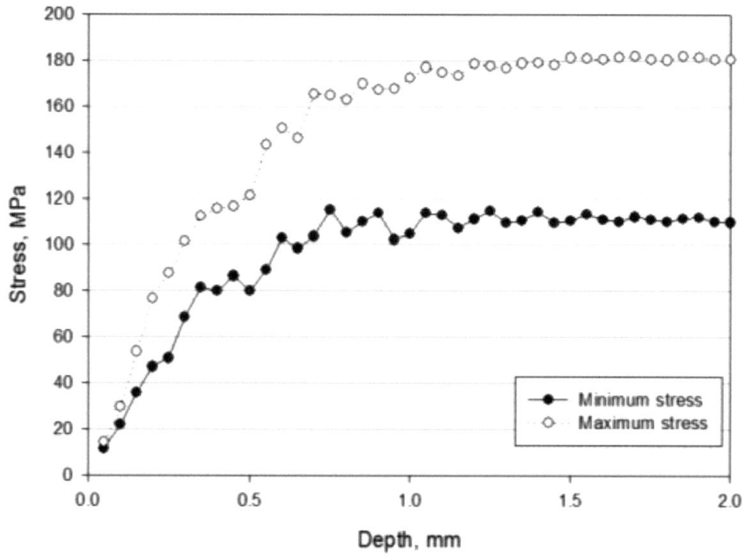

Figure 13.
Laser formed plates (2.5 kW, 90 kJ/m).

velocity is also the slowest in all the speeds used in this study. The slow scanning velocity led to variations in residual stress, when comparing plates irradiated at a power of 26,500 W. Changes in phases associated with the physical properties of the material are related to transformation strains. Strains can be viewed as modes of deformation with the special characteristics of being accompanied by a change in crystal structure [8]. All these factors influence residual stress distribution in titanium. At this power level, there is a reduction in both maximum and minimum stress values, which is in contradiction with other laser powers used in the study. This power had the optimum parameters for a line energy of 90 kJ/m [8] (**Figure 14**).

The specimens processed at 3 kW had a maximum stress of 176 MPa and a minimum stress of 0.3 MPa (C). The residual stresses had a major effect in fatigue

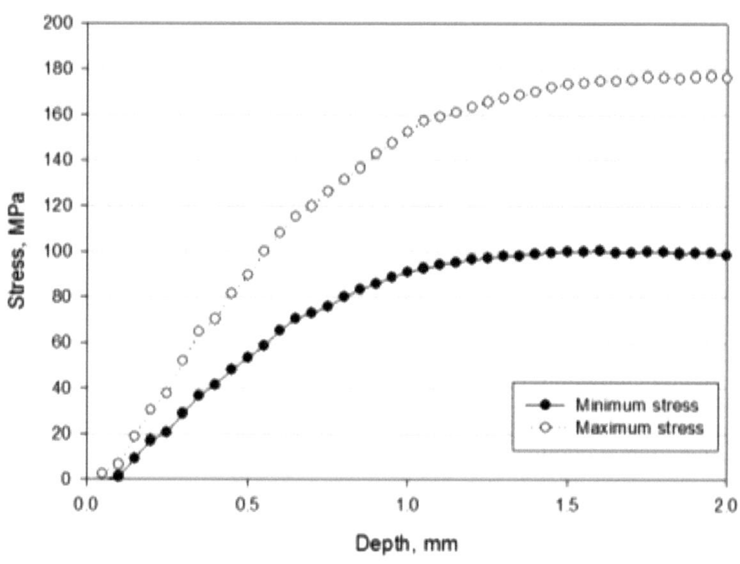

Figure 14.
Laser formed plates (3 kW, 90 kJ/m) [3].

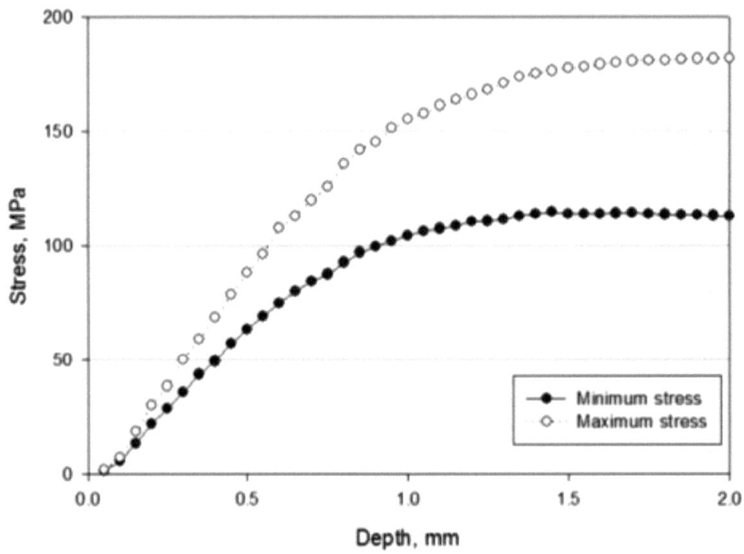

Figure 15.
Relieved stress (3.5 kW).

life as they changed the location of the fracture line. The maximum principal stress at a power of 3000 W is obtained at a depth of 2 mm. The changes in residual stress and strain are closer to the surface of the irradiated plate. The laser forming process increases the hardness of titanium. Residual stresses in this study are a result of interactions between time, temperature and the material. These factors played a major role in the resulting residual stress layout on all laser formed plates. The effect of the thermal gradient is evident when these plates are compared with plates not affected by thermal energy. The highest temperature gradient was obtained at a power of 3500 W. The thermal gradient became a deciding factor in microstructural layout. Even though the line energy was the same from a power of 2500 W up to a power of 3500 W, the effects on the microstructure were not uniform. This led to the conclusion that the thermal gradient is the most influential factor in laser forming [8] (**Figure 15**).

The maximum stress obtained at 3.5 kW was the second highest at 181.9 MPa (T) and a minimum stress of 1.4 MPa (T). The hardness of plates was equivalent to the parent material, but this is where similarities end. The difference in stress was 185 MPa on the laser-processed plates. This difference in stress is related to changes in hardness of titanium as a result of laser forming process. The differences in temperature between 3000 W and 3500 W played no role in influencing minimum and maximum residual stress. The optimum settings for a line energy of **90 kJ/m** are at a power of 2.5 kW [8].

3. Conclusions

The primary motivation of this study was to investigate, analyse, characterise and compare laser and mechanical forming processes. The study focussed on the main parameters that influence the bending of plates and their effect on the microstructure and mechanical properties. New theories and discoveries are discussed in the context of contribution to the subject and body of knowledge which is wide and immense in scope. Theories and conclusions are as follows:

The use of thinner gauge material: The study has come up with a new application for the laser formed titanium. Laser formed titanium plates become extremely hard and could be used in the defence industry for bullet-proof body vests and applied on armoured vehicles. Titanium is light in weight and coupled with a hardened surface resulting from laser forming could be a viable solution. Current armoured vehicles are heavy and slow due to the materials used, and venturing into materials like titanium could be a breakthrough to the defence industry. The laser forming process has the ability to customise the mechanical properties of any material, and therefore with this possibility thinner gauge material could be used for the benefit of this industry.

The control of the radius of curvature: Controlling the radius of curvature using the laser forming process is complex and results in uncontrollable bending of the material. Magee et al. saw a great potential for accuracy and controllability on the amount of forming with the laser, but the current study contradicts what he thought possible with the process. Titanium has proved its unpredictability in this study resulting in no proper control of the radius of curvature as envisaged.

The line energy: This is the fraction of the laser power and traverse speed. According to Magee there is a critical energy input below which no plastic straining occurring in each experiment. The study agrees with Magee on the fact that a higher line energy results in more pronounced bending of the material. Maintaining a constant line energy does not result in same bending of plate specimens from the irradiated batch of plates. An increase in laser power increases the line energy and thermal gradient which all determine the extent of bending in titanium. The use of higher line energies compromises fatigue properties of titanium as there is no proper control of temperature, and therefore precise thermo-mechanical control is needed for the success of this process.

The industrial use of laser forming: Contrary to what was envisaged on initiating this study, the laser forming process does not pose a challenge to current popular forming methods. At the moment the best process for forming is mechanical forming due to the ease with which any desired shape can be formed in minimal time. Laser forming is much slower than mechanical forming, and changes brought by the laser forming process could be undesirable to other industrial applications.

The thermal gradient: The low thermal conductivity of titanium means higher thermal gradients are needed for pronounced bending of CP grade 2 titanium. Higher thermal gradients result in higher residual stresses in the material and a complete change in the physical properties of the material. Changes in physical properties could be desirable or less desirable depending on industrial application.

Surface hardening: The laser forming process resulted in surface hardening of CP grade 2 titanium plates. This had a negative effect on the fatigue life of specimens, as there was a reduction in fatigue life. This is contrary to the findings by Konstantino and Altus [11] who reported improvements in fatigue life of Ti-6Al-4 V which behaves in the same manner as CP grade 2 titanium plates. The improvement in fatigue life was achieved by laser heating based on reduced fraction of α (alpha) in the microstructure and a reduction in grain size. In this study there was an increase in grain size as a result of laser heating which completely changed the granular structure of the material. A significant microstructural refinement was observed during this study resulting in the formation of α-martensite. Hardness values are dependent on the line energy generated during the laser forming of titanium. The higher the line energy, the higher will the hardness be for CP grade 2 titanium plates. The laser-irradiated surface hardens as a result of the laser forming process making it difficult to polish and prepare plates for residual stress measurements. In laser formed CP grade 2 titanium plates, the hardness changes with specimen depth as a result of the effects of thermal energy from the laser, which is the heat source.

The residual stress: In all the forming processes analysed, changes in residual stress are greatly influenced by process specifications. In laser forming however these changes are dependent on the process parameters used, as these differ with each laser power. Thermal gradient influences the development of residual stresses. In mechanical forming changes in residual stress are determined by the complexity of the formed shape. It was envisaged at the beginning of the study that residual stress would be enhanced but the laser forming process made undesirable changes to the underlying residual stress distribution. According to Norton [5] good design requires that an engineer try to tailor the residual stresses to a minimum, not create negative effects on the strength and preferably to create positive effects. Fatigue failure is a tensile residual stress phenomenon. The laser forming process resulted in increased tensile residual stress in the specimens due to higher line energies. The use of lower line energies on CP grade 2 titanium results in no bending of the material, and therefore high tensile residual stress remains part of the process if there is forming to be done. The parent material and mechanically formed plates had low residual stress values which was an advantage during fatigue testing as these had a higher fatigue life.

Microstructure: The mechanical forming process has a minimal influence on the microstructure of CP grade 2 titanium plate specimens compared to the effects of laser forming. The laser forming process results in changes in grain size as thermal energy is increased. Changes in the microstructure influenced the mechanical properties of CP grade 2 titanium plates.

Beam interaction time: The beam interaction time is significant in the analysis of resulting mechanical properties as a result of the laser forming process. The time taken to heat up an area influences the mechanical properties and the microstructure of plate samples.

Heat flux: The heat flux is significant in changes observed with titanium plates, and each laser power setting evaluated had a different reading, resulting in varying microstructures on the plates evaluated. The depth of laser penetration depends on the amount of line energy generated.

Laser power: An increase in laser power leads to the oxidation of the passive layer on CP grade 2 titanium plates as a result of the concentrated thermal energy generated by the laser.

Forming parameters: The changing of forming parameters in the laser forming of CP grade 2 titanium succeeded in obtaining optimum operating parameters (in the case of this research, a power of 2.5 kW, a line energy of **90 kJ/m** and a scanning velocity of 1.67 m/min) for titanium.

Plates: Laser formed titanium plates should not be used in applications requiring prolonged fatigue life. The process is only beneficial in applications where hardness is a priority without a need for high fatigue life, and perhaps the process could be beneficial in military defence applications. Laser formed titanium plates do not bend to the same radius of curvature as proved in this study.

Process control: The process needs precision control of processing parameters as they play a major role in the final microstructural layout and mechanical properties. For good fatigue properties, thermo-mechanical/laser processing of titanium needs to be conducted using lower line energies but then these do not bend the material.

Acknowledgements

Nelson Mandela University (NMMU) for financial assistance and laboratory facilities. Mr. Victor Ngea-Njoume for technical assistance.

Author details

Kadephi Vuyolwethu Mjali[1]* and Annelize Botes[2,3]

1 iThemba LABS—Laboratory for Accelerator Based Sciences, Somerset West, Cape Town, Western Cape, South Africa

2 Council for Scientific and Industrial Research (CSIR), Pretoria, South Africa

3 Department of Mechanical Engineering, Nelson Mandela University, Port Elizabeth, South Africa

*Address all correspondence to: vuyo.mjali@gmail.com

References

[1] Rosi FD, Dube CA, Alexander BH. Mechanisms of plastic flow in titanium-determination of slip and twinning elements. Transactions of AIME. 1953; **197**:257-265

[2] Smith WF. Structure and Properties of Engineering Alloys. 2nd Edition. McGraw-Hill Science/Engineering/Math; January 1 1993. ISBN-13: 978-0070591721

[3] Lawrence J. Advances in Laser Materials Processing. Technology, Research and Applications. 2nd Edition. Woodhead Publishing; ISBN: 9780081012529

[4] Edwardson SP, Watkins KG, Dearden G, French P. Strain Gauge analysis of laser forming. Journal of Laser Applications. 2003;**15**:225. https://doi.org/10.2351/1.1620000

[5] Norton RL. Machine Design-An Integrated Approach. 3rd Edition. Prentice Hall; 2005. ISBN 10: 0131481908/ISBN 13: 9780131481909

[6] Ion JC. Laser processing of engineering materials. Principles, Procedure and Industrial Application. 1st Edition. Butterworth-Heinemann; 2005. ISBN: 978-0-7506-6079-2

[7] Magee J, Watkins KG, Steen WM, Calder N, Sidhu J, Kirby J. Laser forming of high strength alloys. Journal of Laser Applications. 1998;**10**(4):149-155

[8] Mjali KV, Els-Botes A, Mashinini PM. Residual Stress Distribution and the concept of total fatigue stress in laser and mechanically formed commercially pure grade 2 titanium alloy plates ASME 2017 12th International Manufacturing Science and Engineering Conference collocated with the JSME/ASME 2017 6th International Conference on Materials and Processing Volume 2: Additive Manufacturing; Materials. Los Angeles, California, USA, June 4-8, 2017. Conference Sponsors: Manufacturing Engineering Division. ISBN: 978-0-7918-5073-2

[9] Knupfer SM, Moore AJ. The effects of laser forming on the mechanical and metallurgical properties of low carbon and aluminium alloy samples. Materials Science and Engineering A. 25 June 2010;**527**(16-17):4347-4359

[10] Mjali KV, Els-Botes A, Mashinini PM. The Effects of Laser and Mechanical Forming on the hardness and microstructural layout of commercially pure grade 2 titanium alloy plates. ASME 2017 12th International Manufacturing Science and Engineering Conference collocated with the JSME/ASME 2017 6th International Conference on Materials and Processing Volume 2: Additive Manufacturing; Materials. Los Angeles, California, USA, June 4–8, 2017. Conference Sponsors: Manufacturing Engineering Division. ISBN: 978-0-7918-5073-2

[11] Konstantino E, Altus E. Fatigue life enhancement by laser surface treatment. Surface Engineering. 1999; **15**(2):1-3

Section 3

Thermomechanical and Surface Processing

Chapter 4

Processing of Beta Titanium Alloys for Aerospace and Biomedical Applications

Sudhagara Rajan Soundararajan, Jithin Vishnu, Geetha Manivasagam and Nageswara Rao Muktinutalapati

Abstract

The unique combination of attributes—high strength to weight ratio, excellent heat treatability, a high degree of hardenability, and a remarkable hot and cold workability—has made beta titanium alloys an attractive group of materials for several aerospace applications. Titanium alloys, in general, possess a high degree of resistance to biofluid environments; beta titanium alloys with high molybdenum equivalent have low elastic modulus coming close to that of human bone, making them particularly attractive for biomedical applications. Bulk processing of the alloys for aerospace applications is carried out by double vacuum melting followed by hot working. There have been many studies with reference to super-solvus and sub-solvus forging of beta titanium alloys. For alloys with low to medium level of molybdenum equivalent, sub-solvus forging was demonstrated to result in a superior combination of mechanical properties. A number of studies have been carried out in the area of heat treatment of beta titanium alloys. Studies have also been devoted to surface modification of beta titanium alloys. The chapter attempts to review these studies, with emphasis on aerospace and biomedical applications.

Keywords: beta titanium alloys, titanium melting, thermomechanical processing, surface modification, aerospace and biomedical applications

1. Introduction

Titanium was discovered in 1791, but it came into effective application only in the 1950s. After 115 years, i.e., in the year 1906, M. A Hunter at General Electric Company prepared pure titanium for the first time [1]. Since 1950s, titanium holds a prime position in aerospace, biomedical, automotive, and chemical processing industries due to unique features listed below:

1. Low density (60% of steel or super alloy's density),

2. Higher tensile strength (Higher than ferritic stainless steel and comparable to martensitic stainless steel and Fe- base superalloys)

3. Higher operating temperature (Up to 595°C for commercially available alloys and >595°C for titanium aluminides)

4. Excellent corrosion resistance (Higher than stainless steel and biocompatible)

5. Forgeability

6. Castability (Mostly by investment casting)

Despite being the fourth-most abundant structural metal available in the earth crust, its commercial exploitation has been low compared to steel and aluminium due to high cost of production.

Pure Titanium has an hcp crystal structure. Due to the allotropic nature of titanium, the room temperature hcp crystal structure (alpha phase) will be transformed to bcc (beta phase) structure on heating to a particular temperature called beta transus temperature (882.5°C). Alloying elements of titanium are classified on the basis of their influence on the transus temperature. For example, if the transus temperature is increased on the addition of the certain elements, then they are called as alpha stabilisers (Al, O, N, and C); similarly there are some other elements which bring the transus temperature down and they are termed as beta stabilisers (V, Mo, Ta and Nb). The elements Sn and Zr have little or no effect on transus temperature and are termed as neutral elements.

Beta alloys form the metastable beta phase upon quenching rather than undergoing martensitic transformation. A schematic representation of the beta isomorphous phase diagram is shown in the **Figure 1**. Beta alloys can also be classified as those which have alloy which has enough beta stabilisers to avoid the martensitic start (M_s) pass through upon quenching. Beta alloys are further classified into metastable and stable beta alloys based on the content of beta stabilisers. Commercially available beta alloys are metastable beta alloys and stable beta alloys are not commercially available [2]. The metastable beta phase can precipitate the fine alpha

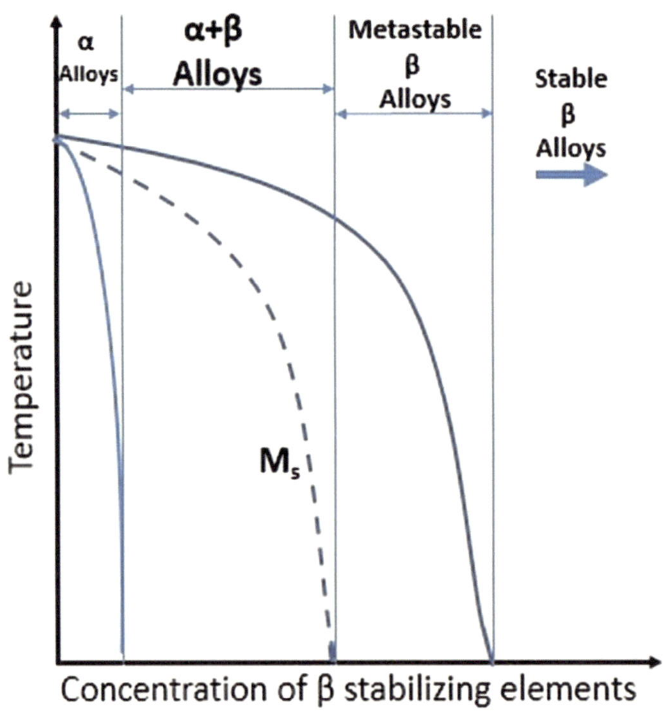

Figure 1.
Beta isomorphous phase diagram.

phase upon ageing/thermal treatment. Hence, beta alloys are hardenable and can attain a higher strength level than alpha + beta alloys and higher specific strength compared to many other alloys [3].

Corrosion resistance of beta alloys is also found to be better than that of alpha + beta alloys. Higher hydrogen tolerance makes beta alloys to perform better in the Hydrogen-rich environments [2]. Increased fracture toughness for a given strength level and amenability to room temperature forming and shaping are superior attributes compared to alpha + beta alloys [1]. Ti-13V-11Cr-3Al (B120VCA) was the first beta alloy produced/developed and used in the SR-71 (Surveillance aircraft) as a sheet product.

Beta alloys' inherent characteristics such as pronounced ductility owing to the crystal structure (bcc), heat treatability, and superior cold rollability make them an effective alternative to alpha + beta alloys [4]. Furthermore, beta alloys have lower beta transus temperature than the alpha + beta alloys [5]. Hence, beta alloys are considered to be the economical choice in perspective of processing compared to the alpha + beta alloys. For example, despite the higher formulation cost, Ti-15V-3Al-3Cr-3Sn alloy's thinner gauges (<2 mm thick) cost one-tenth of those of Ti-6Al-4V [3].

2. Processing of beta alloy

2.1 Melting

The initial step is the fabrication of ingot from sponge for conversion to mill products. The melting practices to produce beta titanium alloy ingots can be broadly categorised into Vaccum Arc Remelting (VAR) and Cold Hearth Melting.

The conventional method used for the melting of beta titanium alloys is the Vacuum Arc Remelting (VAR) in a consumable arc furnace. In VAR, the furnace is initially evacuated for required vacuum and a dc arc is struck between the two electrodes. Here a consumable electrode (material to be melted) is employed as the cathode and starting materials such as titanium-based metal chips or machine turnings act as the anode. The consumable electrode can be fabricated from either of the two strategies.

- From the compacted sponge and/or scrap
- From plasma/electron beam hearth melting

Among these methods, the first method of predensification by compacting using a hydraulic press is widely used to fabricate electrodes. Compacted electrodes with nominal alloy composition are made by the pressing of blended clean and uniform-sized titanium sponge and alloying elements devoid of any harmful inclusions. These compacts (called as briquettes) are then assembled with bulk scrap to form the first melt electrode (called as a stick) by appropriate welding methods.

Finally, these fabricated electrodes are placed inside a vacuum furnace. When the electric arc is established, associated heat generation will result in the dripping of molten metal down to the water-cooled copper crucible to form the ingot. Initially, a layer of solid titanium or skull will be formed on the surface of cooled copper crucible which will hold the subsequently falling molten metal. In order to ensure chemical homogeneity, the ingots will be inverted and remelting will be performed. Ingots produced during first stage melting are again used as consumable electrodes during double or triple remelting. In addition to this, electrical coils are provided in most of the VAR furnaces to generate an electromagnetic field capable of stirring the molten metal thereby further enhancing the homogeneity. Cold hearth melting is another developing technique which uses either plasma arc (PAM) or electron beam (EBM) melting furnace.

Proper monitoring should be ensured to control the solidification of beta titanium based ingots. Specifically, beta eutectoid compositions containing Fe, Mn, Cr, Ni and Cu are associated with depressed freezing temperatures [2]. This allows for solidification over a significant temperature range, consequently leading to solute segregation during solidification of the ingot. Such type of segregation results in regions with lower beta transus and results in a microstructure distinctive from the surrounding material. These solute segregated regions are clearly visible in beta titanium alloys subjected to heat treatment below/near to beta transus and are termed as beta flecks. Beta flecks, which range from a scale of few hundred micrometres to a few millimetres, can act as crack nucleation sites leading to fatigue failure. Beta flecks are mostly developed in large diameter ingots. However, beta isomorphous alloys containing Nb, Mo and V are not associated with these depressed solidification temperatures and are less prone to solute segregation.

Lower values of tensile ductility and low cycle fatigue life of near-β Ti alloy Ti–10V–2Fe–3Al was found to be due to the presence of beta flecks [6]. Under tensile loading, crack nucleation occurred at beta fleck grain boundaries leading to intergranular and quasi-cleavage fracture. In the case of fatigue loading, the inhomogeneous strains developed due to the presence of beta flecks accelerated the crack nucleation and early crack propagation.

2.2 Casting

For an expensive material such as titanium, casting is the perfect choice in attaining a (near) net shape in the fabrication of components with complex geometry without incurring much wastage. A significant weight (35%) saving can be achieved by employing the titanium casting instead of stainless steel casting in B-777 aircraft [7]. In general, rammed graphite mould and investment casting were utilised in titanium casting. Investment casting is preferred to obtain thin sections and better surface finish [8]. Ti-5Al-5V-5Mo-3Cr castings followed by HIP (Hot Isostatic Pressing) possess a superior strength compared to hipped Ti-6Al-4V castings with almost same ductility [9]. To extend brake life of fighter aircraft (F-18 EF) Ti-15V-3Al-3Cr-3Sn castings were used instead of Ti-6Al-4V castings due to the higher specific strength of the former [10].

2.3 Forging and rolling

2.3.1 Ingot breakdown forging

To exploit the ductile nature of the beta phase (bcc crystal structure), even for alpha and alpha + beta alloys, ingot break down forging is done above the beta transus temperature. In general, to avoid thermal stress cracking, titanium alloys are subjected to preheating before high-temperature forging.

Forging is performed to produce billets and bars of titanium with the optimum combination of strength and ductility [11]. Forging is performed using hydraulic presses. Both straight-forging and upset forging are performed in case of Ti alloys. For greater deformation and larger size, upset-forging is preferred [1]. Higher reactivity of the titanium demands the inert / vacuum processing to prevent surface contamination during high-temperature processing [1]. Drawing operation of titanium is prone to galling and seizing. Hence, proper lubricants have to be employed to avoid those effects [1]. Compared to all other Ti alloys, beta alloys can withstand high pressure before cracking. Ti- 13V-11Cr-3Al can withstand up to 690 MPa without cracking. In contrast, Ti-6Al-4V can withstand 585 MPa before cracking [1].

The microstructure of the ingots of beta alloys varies from small equiaxed grains (at the surface) to elongated columnar grains and large equiaxed grains at the bulk/centre of the ingot [4]. Beta Ti alloys are more suitable for low temperature working without being vulnerable to rupture or cracking compared to other Ti alloys [1] and this effect is attributed to the availability of enough slip systems to accommodate the deformations.

2.3.2 Secondary forging

Secondary forging refers to the forging process employed to obtain the final shape/components. The temperature required for this kind of forging is lower than that for ingot breakdown forging. Unlike alpha and alpha + beta alloys, beta alloys show a significant increase in strength at high strain rates [1]. Hence, higher pressures are to be applied for forging of beta alloys; the pressure required to induce crack during forging is higher for beta alloys compared to alpha and alpha + beta alloys [1]. Beta titanium alloys have a broader range of forging temperature compared to alpha/alpha + beta alloys.

Due to the lower beta transus temperature, beta alloys have lower hot working temperature compared to alpha and alpha + beta alloys, For example, Ti–10V–2Fe–3Al has a secondary working temperature range between 700–870°C [12]. Types of forging and features are given in the **Table 1**.

2.3.3 Rolling

Unlike other alloys, rolling of titanium requires higher working pressure and extreme control in temperature. Cylindrical rollers are used to produce the strips, sheet and plate. In contrast, grooved rollers are employed in producing the rounds and other structural shapes. In sheet and plate rolling process, cross rolling is done to reduce the anisotropy in mechanical properties. Texture strengthening is less

S.No.	Forging type	Features
1	Open-Die Forging	• Simple shapes can be made • The previous step to closed – die forging
2	Closed-Die Forging	• It is also called an impression die forging • More complex shapes can be obtained
3	Hot-die forging	• Die is maintained at a higher temperature compared to open and closed die forging • Less energy is required to produce the shape
4	Isothermal Forging	• Preheating the metal is required • Dies are at the same temperature as the metal • Capable of producing near net shape components with less energy required
5	Precision forging	• No machining required before assembling • Highly sophisticated and expensive method • Aero engine fan blades and even aerofoil shapes are precision forged

Table 1.
Types of forging and its features [1].

pronounced in the beta alloys compared to alpha alloys [1]. The lower rate of strain hardening of the beta alloy makes it more acquiescent to cold working.

In Ti-3.5Al-5Mo-6V-3Cr-2Sn-0.5Fe alloy, rolling and ageing in the sub-transus (alpha + beta field) temperature yielded a better combination of the strength and ductility compared to working in the beta field [13]. Sheet beta Ti alloys are amenable to cold rolling. Cold rolling has a strong effect upon mechanical properties. For example, Rosenberg [14] reported the effect of cold rolling on tensile strength, yield strength and ductility of Ti-15-3 alloy:

1. UTS (Rolled alloy) = UTS (un-rolled) + 0.75 × Percentage of reduction (%)

2. YS (Rolled alloy) = YS (un-rolled) + 0.65 × Percentage of reduction (%)

3. Ductility (Rolled alloy) = EL (un-rolled) – 0.65 × Percentage of reduction (%)

Two high roll mill and three high roll mill are commonly used for rolling titanium and its alloys.

2.4 Thermomechanical processing

Material processing performed with the aid of both mechanical force and thermal/ heat treatment can be termed as thermomechanical processing. The primary objective of this processing is to obtain a component in functional design with pre-determined microstructure and corresponding mechanical properties. Thermomechanical processing of beta Ti alloys can be done both above transus temperature (Super-transus processing) and below the transus temperature (Sub-transus processing). Super-transus processing with hot deformation is optimised to obtain fine recrystallised beta grains. Sub-transus processing is optimised to obtain fine beta grains with controlled alpha phase morphology [12]. Size, volume fraction, morphology, and the spatial distribution of the alpha precipitates formed during the thermomechanical processing have a vital influence over the mechanical properties of the end product.

In Ti-15V-3Al-3Cr-3Sn alloy, Boyer et al. [15], showed the usefulness of thermomechanical treatment for attaining a wide range of tensile strength (from 1070 to 1610 MPa.)

2.5 Heat treatment

Heat treatment is the basic metallurgical process through which optimization of hardness, tensile strength, fatigue strength and fracture toughness can be achieved. All the metastable beta alloys are heat treatable to attain higher strength than alpha + beta alloys.

Duplex ageing treatment yielded a superior combination of mechanical properties with no precipitation free zone and finer alpha precipitation compared to single ageing in Ti-15V-3Al-3Cr-3Sn-3Zr [16] and Ti-3Al-8V-6Cr-4Mo-4Zr [17]. The rate of heating to ageing temperature was found to have a substantial effect on the evolution of microstructure and mechanical properties [18]. Choice of solution treatment temperature is important. For example, for Ti-1Al-8V-5Fe (Ti185), solution treatment near beta transus temperature leads to a highest tensile and yield strength [19].

Solution treatment followed by ageing in metastable beta alloys will lead to a microstructure consisting of soft alpha in the beta grain boundaries. Hence, this softer alpha phase may lead to the decline in the HCF behaviour [20] and tensile ductility by augmenting the intergranular fracture [17]. For example, Sauer and

Luetjering [21] have also reported the adverse effect of alpha phase layers along the beta grain boundaries on the tensile and fatigue behaviour of Ti-5Al-2Sn-4Zr-4Mo-2Cr-1Fe (β CEZ).

2.6 Surface processing for aerospace application

Modifying the surface is an effective and economical way to enhance the tribological and fatigue properties of the material. Thermo-chemical and mechanical surface modification techniques are common in beta alloys.

2.6.1 Thermo-chemical surface modification

In order to enhance the surface hardness, wear resistance and near-surface strength, thermo-chemical surface processing techniques such as nitriding and carburising are employed. Among various thermo-chemical surface processing techniques, nitriding is extensively used. In this process, the nitrogen is fused into the titanium base alloy. Among the various technologies used for Nitriding, i.e., gas nitriding, laser nitriding, plasma nitriding, Ion nitriding and gas Nitriding are used widely [22]. Titanium nitrides will be formed on the surface as a result of the nitriding and these nitrides increase the surface hardness drastically and improve the tribological properties at the expense of the ductility of the material. Increased hardness due to TiN formation was made use in flap tracks of Military airplanes [23]. However, nitriding has a negative influence on the tensile strength and fatigue strength of the material.

2.6.2 Mechanical surface modification

Mechanical surface modifications such as shot peening, ball burnishing and laser peening are developed to enhance the fatigue behaviour of the target material by inducing the residual compressive stress and work hardening effect in near surface region. Both crack nucleation and crack propagation during fatigue loading were found to be affected by the surface modification treatment. However, surface roughness will be significantly increased at the end of the mechanical surface modification such as shot peening and this may lead to early crack initiation.

Since 1970s, shot peening is being employed in enhancing the mechanical behaviour of Ti alloys in aerospace industries [24]. Schematic representation of shot peening is shown in the **Figure 2**. Shot peening of beta alloys, i.e. Ti-10V-2Fe-3Al and Ti-3Al-8V-6Cr-4Mo-4Zr yielded a marginal increase in the fatigue life compared to electro polished sample [25]. In LCB beta alloy, in order to compensate the residual compressive stress induced in the surface after peening, substantial tensile residual stress formed in the subsurface region and this deteriorated the fatigue behaviour compared to polished sample [26]. It is important to control the shot peening conditions to get the desired enhancement in fatigue life.

Unlike shot peening and laser peening, roller burnishing reduces the surface roughness by stressing the surface with a roller ball with optimised pressure. Schematic representation of the roller burnishing is shown in the **Figure 3**. Roller burnishing of Ti-10V-2Fe-3Al beta alloy induced deeper and higher magnitude residual stress compared to shot peening. In roller burnishing of LCB beta alloy, higher the rolling pressure, deeper was the site for fatigue crack nucleation [27]. In Beta C (Ti-3Al-8V-6Cr-4Mo-4Zr) alloy, deep rolling ended up with deeper residual stress distribution compared to shot peening, but the magnitude of the residual stress remained high for the shot peened sample. A marginal increase in fatigue life was achieved through deep rolling of Beta C alloy [28].

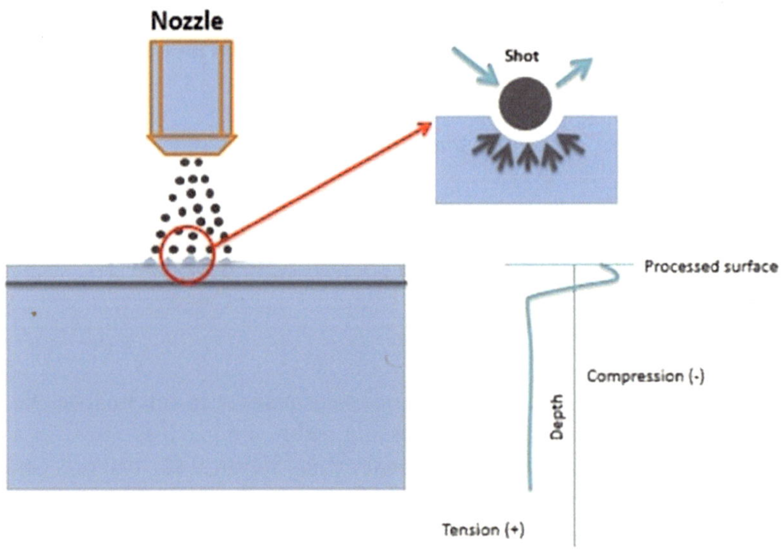

Figure 2.
Schematic representation of shot peening.

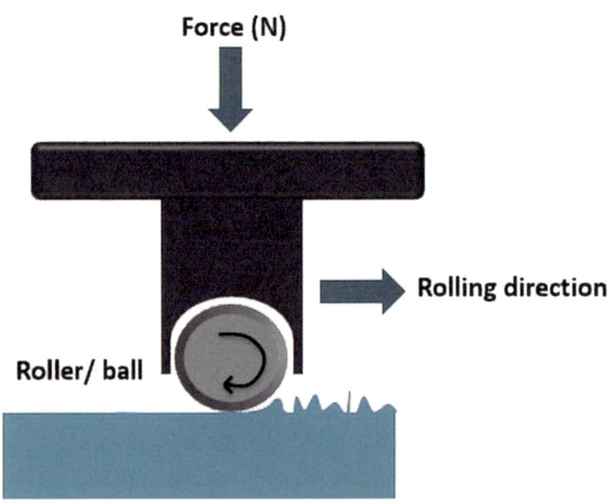

Figure 3.
Schematic representation of the ball burnishing.

Compared to shot peening, laser peening has unique features like the capability of inducing deeper and stable residual stress with extreme control in operation. Conventionally, laser peening is performed using Nd: Glass lasers after applying the coating, i.e. black paint on the target surface. To make this process simple, economical and more portable, LPwC (Laser peening without Coating) was developed in 1995 [29]. LPwC has proven to be an effective technique by inducing a relatively high compressive residual stress. For example, a residual stress of approx. −825 MPa was induced at a depth of ~75 μm from the surface in LCB (Ti-6.8Mo-4.5Fe-1.6Al) beta alloy [30].

2.7 Surface processing for bio-medical application

In the case of implant materials, the interaction between the biological environment and the implanted materials occurs on the biomaterial surface. Clinical success of implant materials is greatly dependent on various surface characteristics viz. chemical inertness, texture, corrosion resistance and surface energy [31] . In the case of orthopaedic implants, the surface should possess more bone forming ability and for blood contacting devices, it should not initiate any blood clot formation. Hence surface modification of biomedical grade beta titanium alloys is very significant. Oxide layer formation will occur spontaneously on the surface of titanium on exposure to air. This TiO_2 film possesses a thickness of about 1.5 to 10 nm at room temperature. Chemical stability and structural characteristics of this oxide film greatly influence the biocompatibility of titanium implant materials. Some of the potential methods to enhance the properties of native TiO_2 film are anodisation, sol–gel methods, acidic and alkaline treatments [32]. In addition to these, specific surface topographies and roughness induced by mechanical surface modifications (sandblasting, grit blasting, peening) have improved the clinical success of implant materials. An overview of the various surface modification techniques employed for biomedical beta titanium alloys is schematically shown in **Figure 4**.

In dental applications, Laser Nitriding has proved to be an effective process in enhancing the surface hardness, the coefficient of friction and corrosion resistance of the Ti-20Nb-13Zr and wear and corrosion resistance of Ti-13Nb-13Zr biomedical-beta alloys [34, 35]. Plasma nitrided beta 21S (Ti-15Mo-3Nb-3Al-0.2Si) alloy showed higher hardness but inferior corrosion resistance compared to the untreated alloy [36]. In line with the Nitriding, carburising of Ti-13Nb-13Zr (a biomedical beta alloy used for artificial joints) improved the surface hardness and wear resistance through the formation of the titanium carbide [37].

2.8 Powder metallurgy

As mentioned in the introduction (Section 1), a major limiting factor for the titanium application is its high production cost. In addition to the high raw

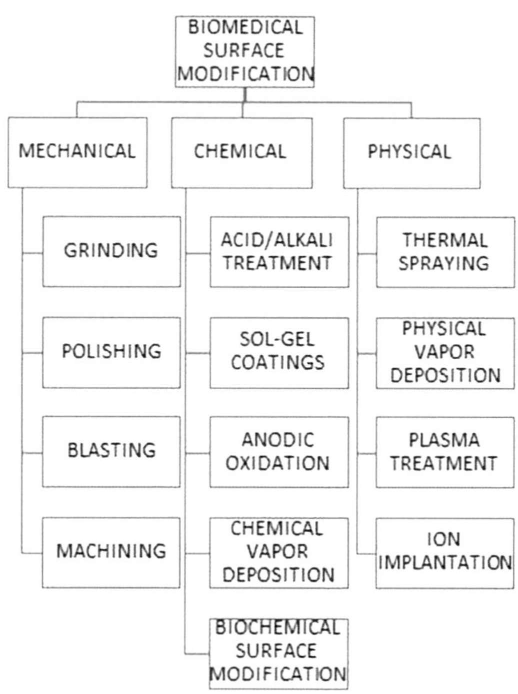

Figure 4.
Overview of surface modification of beta titanium alloys for biomedical application [33].

material cost, the forging, machining contribute majorly to the production cost. This limitation instigated the industries to work towards processing methods through which the near net shape (NNS) could be obtained. Despite the higher cost involved, Powder metallurgy of titanium is capable of yielding almost same or better mechanical properties compared to wrought and cast components along with accurate net shape capability. This merit is mainly attributed to the absence of texture, segregation and nonuniformity in the grain size encountered in conventional processing.

Even for the components made through powder metallurgy route, solution treatment followed by ageing (STA) leads to an enhancement in mechanical properties such as tensile strength and yield strength compared to the as-sintered condition [38]. Ti-10V-2Fe-3Al and Ti-11.5Mo-6Zr-4.5Sn alloys have been produced through powder metallurgy route. However, 90% of the powder metallurgy is focussed on the alpha + beta alloy Ti-6Al-4V.

Guo et al. [39] reported a remarkable increase in the mechanical properties of Ti-10V-2Fe-3Al powder alloy compared to the wrought and cast products through isothermal forging of the sintered alloy. Jiao et al. [40] studied the model of alpha phase spatial distribution in laser additive manufactured Ti-10V-2Fe-3Al. The influence of nano-scale alpha precipitates on tensile properties of age hardened laser additive manufactured Ti-5Al-5Mo-5V-1Cr-1Fe (Ti-55,511) alloy was studied by He et al. [41] and the authors reported that precipitated nanoscale alpha precipitates have led to a decline in ductility.

3. Applications

3.1 Aerospace applications

A recent forecast released by Airbus Industries [42], confirms the promising development of air transport requiring 37,400 aircraft at a value of 5.8 trillion US dollars business in the next 20 years. However, reducing the fuel consumption to control the emission of CO_2 and NO_x is the driving factor for the aerospace industries and this could be possible by reducing the overall weight [43]. Similarly, in space application weight of the payload is more crucial than civil/cargo aviation. Ti-6Al-4V is a workhorse for the aerospace industry for several decades and 65% of total titanium production in the United States belongs to Ti-6Al-4V alloy [3].Even though the alpha +beta alloys dominated the scene, beta alloys with their unique characteristics such as excellent hardenability, heat treatability to high strength levels and a high degree of sheet formability, are becoming increasingly important for the aerospace sector. Beta alloys and their aerospace application are listed in the **Table 2**.

3.2 Biomedical applications

Titanium is the ultimate choice for biomedical applications as they outperform conventionally used biomedical alloys such as 316L stainless steel and cobalt-chromium alloys [47]. The formation of a nanometre thick oxide layer on titanium when exposed to any environment imparts high corrosion resistance and superior biocompatibility [48]. All classes of titanium α, $\alpha + \beta$, near β and β alloys are widely used for biomedical applications.

Despite being initially developed for aerospace applications, CP titanium and Ti-6Al-4V are still the most widely used Ti grades being used for biomedical applications. However, CP Ti is associated with lower wear resistance and Ti-6Al-4V when implanted inside the body releases Al and V ions which can

S. No.	Alloy	Application/components
1	Ti-15V-3Al-3Cr-3Sn	Landing gear, springs, sheet, plate and airframe castings, environmental control system ducting
2	Ti-6V-6Mo-5.7Fe-2.7Al	Fasteners
3	Ti-13V-11Cr-3Al	Airframe, landing gear and springs
4	Ti-3Al-8V-6Cr-4Mo-4Zr (β-C)	Springs and fasteners
5	Ti-11.5Mo-6Zr-4.5Sn	Rivbolts—Boeing 747
6	Ti-5Al-5Mo-5V-3Cr	Aircraft landing gear, Fuselage components and high lift devices
7	Ti-10V-2Fe-3Al	1. Aircraft landing gear 2. High strength forging
8	Beta 21s	1. Nozzle assembly parts in Boeing 777 2. Planned to use in Pratt & Whitney PW4168 engine components
9	Ti-35V-15Cr (Alloy C)	Compressor and exhaust nozzle components

Table 2.
Aerospace applications of beta titanium alloys [4, 9, 17, 43–46].

lead to severe neurological disorders and allergic reactions. Moreover, the elastic modulus values of these alloys (~110 GPa) are almost four times than that of human cortical bone (20–30 GPa) which can lead to stress shielding effect. This led to the development of β-Ti alloys composed of non-toxic elements and their inherent lower elastic modulus assists in reducing the stress shielding effect when used for orthopaedic applications [3]. Alloy systems based on Ti-Nb, Ti-Mo, Ti-Ta and Ti-Zr are potential materials for biomedical applications. Some of these β-Ti alloys initially developed are Ti-15Mo-5Zr-3Al, Ti-12Mo-6Zr-2Fe (TMZF), Ti-15Mo-3Nb-0.3O (21SRx) and Ti-13Nb-13Zr possessing modulus values in the range of 70–90 GPa.

In the early 1990s, medical device industry focused on developing these low modulus β-Ti alloys for orthopaedic applications. Initially, two β-Ti alloys Ti-13Nb-13Zr specified by ASTM F1713 and Ti-12Mo-6Zr-2Fe (TMZF) specified by ASTM F1813 received Food and Drug Administration approval as implant materials. Among these, TMZF alloy possesses an elastic modulus of about 74–85 GPa, with a yield strength of 1000 MPa. During the early 2000s, this metastable β-Ti alloy was used for making hip stems, which rub against a modular neck made from a cobalt-chromium based alloy. However, in 2011, the US Food and Drug Administration recalled the use of this TMZF alloy due to the unacceptable level of wear debris formation. Another β-Ti alloy 21SRx is derived from the aerospace alloy 21S from which aluminium was eliminated over biocompatibility concerns. In addition, alloys such as Ti-29Nb-13Ta-4.6Zr and Ti-35Nb-7Zr-5Ta are receiving increasing attention due to their lower elastic moduli of about 65 and 55 GPa, respectively, lower than other β-Ti alloys [50].

Apart from orthopaedics, titanium is extensively used in the dental applications [49]. In the case of orthodontic wire material, it should possess three general characteristics viz. large spring back (ability to be deflected over longer distances without permanent deformation), lower stiffness and high formability [51]. The initially utilised materials for orthodontic wire application were gold based alloys containing copper, palladium, platinum or nickel. However, spring back values of these gold alloys were limited owing to their lower yield strength. In the 1960s gold was replaced by stainless steel and cobalt-chromium based alloy (elgiloy). These materials continue to be the standard orthodontic wire material for the past

70 years and possess higher springiness and strength with comparable corrosion resistance. During the early 1970s, nickel-titanium alloy Nitinol (Nickel Titanium Naval Ordinance Laboratory) was also used for orthodontic wires. Even though Nitinol orthodontic archwires are widely used owing to their superior superelastic properties, their use is hampered by reduced formability during the final stages of treatment. Moreover, there are serious concerns over the nickel ion release from these materials in the oral environment. It was later demonstrated that orthodontic wires made from β-Ti alloy Ti-11.3Mo-6.6Zr-4.3Sn (TMA alloy) possess enhanced spring back and formability, along with reduced stiffness. TMA alloys possess ideal elastic modulus values lower than that of stainless steels and higher than nitinol [51]. The higher surface roughness associated with these TMA wires can, however, lead to arch wire-bracket sliding friction due to the high coefficient of friction of TMA alloys. One of the most successful approaches to tackle this problem is the ion implantation process which renders the TMA wires with lower surface roughness and reduced friction coefficients. Another beta titanium alloy Ti-6Mo-4Sn was also investigated for orthodontic wire applications. By proper heat treatment procedures, this alloy exhibited an elastic modulus of 75 GPa and a tensile strength of 1650 MPa [52]. Ti-13V-11Cr-3Al, metastable Ti-3Al-8V-6Cr-4Mo-4Zr, metastable Ti-15V-3Cr-3Al-3Sn, near-beta Ti-10V-2Fe-3Al were also researched for dental archwire applications.

Though beta titanium alloys possess superior haemocompatibility, which is beneficial for cardiovascular devices, they are not fully exploited for cardiovascular applications. Despite higher haemocompatibility, no β-Ti alloy based stents have been commercialised which can be attributed to their lower ductility and modulus as compared to 316L stainless steel and cobalt-chromium based stent materials. Recently, research based on the development of new β-Ti alloy compositions for coronary stent applications has been getting increased attention. Initial studies on Ti-12Mo (wt %) and ternary Ti-9Mo-6W (wt %) demonstrated a ductility of about 46% and 43% respectively [53]. Apart from this, initial investigations on Ti-50Ta, Ti-45Ta-5Ir and Ti-17Ir for stent applications were performed by Brien et al. [54]. Among the three alloys, Ti-17Ir exhibited a favourable elastic modulus of 128 GPa owing to the eutectoid Ti3Ir phase precipitation; iridium content will also assist in improving the fluoroscopic visibility of the stents during interventional procedures [54].

4. Conclusions

Beta titanium alloys have shown much promise and extensive research and development work has been devoted to this group of alloys over the last four decades. For aerospace applications, their heat treatability, high hardenability, high strength to weight ratio and excellent hot and cold workability are major attractions. For orthopaedic applications, their corrosion resistance to biofluids, biocompatibility and low elastic modulus coming close to that of human bone are the important attractive features. Accordingly, development of cost-effective processing techniques has also assumed importance. Problems unique to beta titanium alloys such as high degree of proneness to segregation, high loads to be applied during hot working etc. have since been resolved. Powder processing and additive manufacturing of the alloys have recently received attention and hold promise. Surface modification has been an important part of the developmental efforts and has taken a prominent place, especially for biomedical applications. Coming years are bound to witness increased exploitation of this group of alloys, particularly in biomedical and aerospace applications.

Acknowledgements

The authors would like to express their gratitude to the Management of Vellore Institute of Technology (VIT)—Vellore campus, Tamil Nadu, India, for allowing us to submit this manuscript.

Conflict of interest

The authors declare that there is no conflict of interest regarding the publication of this article.

Author details

Sudhagara Rajan Soundararajan[1], Jithin Vishnu[2], Geetha Manivasagam[1,2] and Nageswara Rao Muktinutalapati[1*]

1 School of Mechanical Engineering, Vellore Institute of Technology (VIT), Vellore, Tamil Nadu, India

2 Centre for Biomaterials, Cellular and Molecular Theranostics, Vellore Institute of Technology (VIT), Vellore, Tamil Nadu, India

*Address all correspondence to: muktinutala@gmail.com

References

[1] Froes FH. Titanium: Physical Metallurgy Processing and Application. Materials Park, Ohio: ASM International; 2014

[2] Lütjering G, Williams JC. Titanium. 2nd ed. Berlin, Germany: Springer; 2007

[3] Bania PJ. Beta titanium alloys and their role in the titanium industry. Journal of Metals. 1994;**46**:16-19. DOI: 10.1007/BF03220742

[4] Froes FH, Bomberger HB. The Beta titanium alloys. Journal of Metals. 1985;**37**:28-37. DOI: 10.1007/BF03259693

[5] Kolli R, Devaraj A. A review of metastable Beta titanium alloys. Metals (Basel). 2018;**8**:506. DOI: 10.3390/met8070506

[6] Zeng WD, Zhou YG. Effect of beta flecks on mechanical properties of Ti–10V–2Fe–3Al alloy. Materials Science and Engineering A. 1999;**260**:203-211. DOI: 10.1016/S0921-5093(98)00954-X

[7] Boyer RR. Aerospace applications of beta titanium alloys. Journal of Metals. 1994;**46**:20-23. DOI: 10.1007/BF03220743

[8] Leyens C, Peters M. Titanium wand Titanium Alloys Fundamentals and Applications. Weinheim: Wiley-VCH Verlag GmbH & Co. KGaA; 2003. DOI: 10.1002/3527602119

[9] Boyer RR, Briggs RD. The use of beta titanium alloys in the aerospace industry. Journal of Materials Engineering and Performance. 2005;**14**:680-684. DOI: 10.1007/s11665-013-0728-3

[10] Boyer RR. An overview on the use of titanium in the aerospace industry. Materials Science and Engineering A. 1996;**213**:103-114. DOI: 10.1016/0921-5093(96)10233-1

[11] Donachie M. Introduction to selection of titanium alloys. In: Titanium: A Technical Guide. Materials Park, Ohio: ASM International; 2000. pp. 5-11

[12] Weiss I, Semiatin SL. Thermomechanical Processing of Beta Titanium Alloys—An Overview. Materials Science and Engineering A. Elsevier; 1998. DOI: 10.1016/S0921-5093(97)00783-1

[13] Du ZX, Xiao SL, Shen YP, Liu JS, Liu J, Xu LJ, et al. Effect of hot rolling and heat treatment on microstructure and tensile properties of high strength beta titanium alloy sheets. Materials Science and Engineering A. 2015;**631**:67-74. DOI: 10.1016/j.msea.2015.02.030

[14] Rosenberg HW. Ti-15-3: A new cold-formable sheet titanium alloy. Journal of the Minerals, Metals and Materials Society. 1983;**35**:30-34. DOI: 10.1007/BF03339165

[15] Boyer R, Rack H, Venkatesh V. The influence of thermomechanical processing on the smooth fatigue properties of Ti–15V–3Cr–3Al–3Sn. Materials Science and Engineering A. 1998;**243**:97-102. DOI: 10.1016/S0921-5093(97)00785-5

[16] Santhosh R, Geetha M, Saxena VK, Nageswara Rao M. Effect of duplex aging on microstructure and mechanical behavior of beta titanium alloy Ti-15V-3Cr-3Al-3Sn under unidirectional and cyclic loading conditions. International Journal of Fatigue. 2015;**73**:88-97. DOI: 10.1016/j.ijfatigue.2014.12.005

[17] Santhosh R, Geetha M, Rao MN. Recent developments in heat treatment of Beta titanium alloys for aerospace applications. Transactions of the Indian Institute of Metals. 2016. DOI: 10.1007/s12666-016-0985-6

[18] Santhosh R, Geetha M, Saxena VK, Nageswararao M. Studies on single

and duplex aging of metastable beta titanium alloy Ti-15V-3Cr-3Al-3Sn. Journal of Alloys and Compounds. 2014;**605**:222-229. DOI: 10.1016/j.jallcom.2014.03.183

[19] Devaraj A, Joshi VV, Srivastava A, Manandhar S, Moxson V, Duz VA, et al. A low-cost hierarchical nanostructured beta-titanium alloy with high strength. Nature Communications. 2016;**7**:1-8. DOI: 10.1038/ncomms11176

[20] Benjamin R, Nageswara Rao M. Crack nucleation in β titanium alloys under high cycle fatigue conditions - a review. Journal of Physics: Conference Series. 2017;**843**. DOI: 10.1088/1742-6596/843/1/012048

[21] Sauer C, Luetjering G. Thermo-mechanical processing of high strength β- titanium alloys and effects on microstructure and properties. Journal of Materials Processing Technology. 2001;**117**:311-317. DOI: 10.1016/S0924-0136(01)00788-9

[22] Sha W, Malinov S. Titanium Alloys: Modelling of Microstructure, Properties and Applications. Abington Hall, Granta Park, Great Abington, Cambridge CB21 6AH, UK: Woodhead Publishing Limited; 2009

[23] Weisheit A, Mordike BL. Laser nitriding of titanium alloys. Developments in Nitriding Iron Titananium Based Alloy. 1995:145-170. DOI: 10.1179/sur.1995.11.2.145

[24] Lain SJ, Knowles KM, Doorbar PJ, Cutts RD, Rugg D. Microstructural characterisation of metallic shot peened and laser shock peened Ti-6Al-4V. Acta Materialia. 2017;**123**:350-361. DOI: 10.1016/j.actamat.2016.10.044

[25] Dörr T, Wagner L. Fatigue response of various titanium alloys to shot peening. Transactions on Engineering Sciences. 1999;**25**:589-598

[26] Kocan M, Rack HJ, Wagner L. Fatigue performance of metastable beta titanium alloys: Effects of microstructure and surface finish. Journal of Materials Engineering and Performance. 2005;**14**:765-772. DOI: 10.1361/105994905x75583

[27] Ludian T, Kocan M, Rack HJ, Wagner L. Residual-stress-induced subsurface crack nucleation in titanium alloys. International Journal of Materials Research. 2006;**97**:1425-1431

[28] Wagner L. Mechanical surface treatments on titanium, aluminum and magnesium alloys. Materials Science and Engineering A. 1999;**263**:210-216. DOI: 10.1016/S0921-5093(98)01168-X

[29] Mukai N, Aoki N, Obata M, Ito A, Sano Y, Konagai C. Laser processing for underwater maintenance in nuclear plants. 3rd JSME/ASME Joint International Conference on Nuclear Engineering. 1995;**3**:1489-1494

[30] Maawad E, Sano Y, Wagner L, Brokmeier HG, Genzel C. Investigation of laser shock peening effects on residual stress state and fatigue performance of titanium alloys. Materials Science and Engineering A. 2012;**536**:82-91. DOI: 10.1016/j.msea.2011.12.072

[31] Liu X, Chu PK, Ding C. Surface modification of titanium, titanium alloys, and related materials for biomedical applications. Materials Science & Engineering R: Reports. 2004;**47**:49-121. DOI: 10.1016/j.mser.2004.11.001

[32] Ou HH, Lo SL. Review of titania nanotubes synthesized via the hydrothermal treatment: Fabrication, modification, and application. Separation and Purification Technology. 2007;**58**:179-191. DOI: 10.1016/j.seppur.2007.07.017

[33] Liu X, Chu PK, Ding C. Surface modification of titanium, titanium alloys, and related materials for biomedical applications. Materials Science & Engineering R: Reports. 2004;**47**:49-121. DOI: 10.1016/j.mser.2004.11.001

[34] Hussein MA, Yilbas B, Kumar AM, Drew R, Al-Aqeeli N. Influence of laser Nitriding on the surface and corrosion properties of Ti-20Nb-13Zr alloy in artificial saliva for dental applications. Journal of Materials Engineering and Performance. 2018. DOI: 10.1007/s11665-018-3569-2

[35] Sathish S, Geetha M, Pandey ND, Richard C, Asokamani R. Studies on the corrosion and wear behavior of the laser nitrided biomedical titanium and its alloys. Materials Science and Engineering: C. 2010;**30**:376-382. DOI: 10.1016/j.msec.2009.12.004

[36] Mohan L, Anandan C. Effect of gas composition on corrosion behavior and growth of apatite on plasma nitrided titanium alloy Beta-21S. Applied Surface Science. 2013;**268**:288-296. DOI: 10.1016/j.apsusc.2012.12.080

[37] Luo Y, Rao X, Yang T, Zhu J. Effect of solid carburization on the Tribological behaviors of Ti13Nb13Zr alloy. Journal of Tribology. 2017;**140**. DOI: 031604, 10.1115/1.4038269

[38] Donachie MJ. Titanium: A Technical Guide. Materials Park, Ohio: ASM International; 2000. DOI: 10.5772/1844

[39] Guo H, Zhao Z, Duan C, Yao Z. The Powder Sintering and Isothermal Forging of Ti-10V-2Fe-3Al. Journal of the Minerals, Metals and Materials Society. 2008;**60**(11):47-49

[40] Jiao Z, Fu J, Li Z, Cheng X, Tang H, Wang H. The spatial distribution of α phase in laser melting deposition additive manufactured Ti-10V-2Fe-3Al alloy. Materials and Design. 2018;**154**:108-116. DOI: 10.1016/j.matdes.2018.05.032

[41] He B, Li J, Cheng X, Wang HM. Brittle fracture behavior of a laser additive manufactured near-β titanium alloy after low temperature aging. Materials Science and Engineering A. 2017;**699**:229-238. DOI: 10.1016/j.msea.2017.05.050

[42] Airbus. Global Market Forecast. 2018;**2018-2037**:1-2. https://www.airbus.com/aircraft/market/global-market-forecast.html

[43] Antunes RA, de Oliveiraa MCL, Salvador CAF. Materials selection of optimized titanium alloys for aircraft applications. Materials Research. 2018;**21**. DOI: 10.1590/1980-5373-mr-2017-0979

[44] Peters M, Kumpfert J, Ward CH, Leyens C. Titanium alloys for aerospace applications. Advanced Engineering Materials. 2003;**5**:419-427. DOI: 10.1002/adem.200310095

[45] Zhang X, Chen Y, Hu J. Recent advances in the development of aerospace materials. Progress in Aerospace Science. 2018;**97**:35-60. DOI: 10.1016/j.paerosci.2018.01.001

[46] Cotton JD, Briggs RD, Boyer RR, Tamirisakandala S, Russo P, Shchetnikov N, et al. State of the art in Beta titanium alloys for airframe applications. Journal of Metals. 2015;**67**:1281-1303. DOI: 10.1007/s11837-015-1442-4

[47] Geetha M, Singh AK, Asokamani R, Gogia AK. Ti based biomaterials, the ultimate choice for orthopaedic implants - a review. Progress in Materials Science. 2009;**54**:397-425. DOI: 10.1016/j.pmatsci.2008.06.004

[48] Dhinasekaran D, Geetha Manivasagam AR. Biomedical

implants: Corrosion and its prevention - a review. Recent Patents on Corrosion Science. 2010;**2**:40-54. file:///C:/Users/Wissha/Documents/EUETIB UNIVERSIDAD 2010-2011/9° CUATRIMESTRE (FINAL)/TFG/varios/RPTCS-2-40.pdf

[49] Revathi A, Borrás AD, Muñoz AI, Richard C, Manivasagam G. Degradation mechanisms and future challenges of titanium and its alloys for dental implant applications in oral environment. Materials Science and Engineering: C. 2017;**76**:1354-1368. DOI: 10.1016/j.msec.2017.02.159

[50] Rack HJ, Qazi JI. Titanium alloys for biomedical applications. Materials Science and Engineering: C. 2006;**26**:1269-1277. DOI: 10.1016/j.msec.2005.08.032

[51] Burstone CJ, Goldberg AJ. Beta titanium : Alloy. American Journal of Orthodontics. 1980;**77**:121-132

[52] Hida M, Miyazawa K, Tsuruta S, Kurosawa M, Hata Y, Kawai T, et al. Effect of heat treatment conditions on the mechanical properties of Ti-6Mo-4Sn alloy for orthodontic wires. Dental Materials Journal. 2013;**32**:462-467. DOI: 10.4012/dmj.2012-118

[53] Prima F, Sun F, Vermaut P, Gloriant T, Mantovani D, Jacques PJ. High performance Beta titanium alloys as a new material perspective for cardiovascular applications. Materials Science Forum. 2012;**706-709**:578-583. DOI: 10.4028/www.scientific.net/MSF.706-709.578

[54] O'Brien B, Stinson J, Carroll W. Initial exploration of Ti-Ta, Ti-Ta-Ir and Ti-Ir alloys: Candidate materials for coronary stents. Acta Biomaterialia. 2008;**4**:1553-1559. DOI: 10.1016/j.actbio.2008.03.002

Chapter 5

Characteristics of the Dissipation of Energy at Hot Plastic Deformation of Near-Alpha Titanium Alloy

Mikhail Mikhaylovich Radkevich,
Nikolay Rafailovich Vargasov
and Boris Konstantinovich Barakhtin

Abstract

Change of mechanical properties of near-alpha titanium alloy is experimentally investigated at stretching in the conditions of variation of temperature and high-speed parameters of deformation. It is established that characteristics of mechanical properties, a structural state influence processes of dissipation of the spent energy. Studying of microstructure of samples before deformation by stretching allowed to install the main mechanisms of dissipative processes and to confirm a possibility of realization of superplasticity in the studied alloy.

Keywords: titanium alloy, tensile deformation, dissipation, superplasticity, microstructure

1. Introduction

Structural and phase transformations in metal alloys at deformation in the conditions of plasticity and superplasticity are a subject of long-term and systematic researches.

In scientific literature there are physical and mathematical models of deformation describing structural transformations in process as plastic and superplastic deformation of structural materials [1, 2].

However we have very few materials of publications in which results of the thermodynamic analysis directly correspond to researches of structural transformations. Authors of works [3–5] showed that one of the effective methods of studying of mechanisms of hot plastic deformation is the thermodynamic approach based on use of dynamic model of deformation of material.

According to the model of the elasto-visco-plastic environment, for any time-point the power of the mechanical energy P coming to a deformable body is defined by the sum composed by G and J. Both are connected with production of entropy. However first (G) considers dissipation of energy through forming and hardening. Second (J) is connected with the adapting reorganizations in structure of grains of a polycrystal directly in the course of action of the deforming tension. Hence it is connected with production of entropy in material:

$$P = G + J = \sigma\varepsilon; = T(ds/dt) \geq 0, \tag{1}$$

where σ is tension, ε; is strain rate, T is temperature, and dS/dt is the speed of production of entropy.

Division of power of dissipation between G and J is defined by strain rate sensitivity m:

$$dJ/dG = \Delta\log(\sigma)/\Delta\log(\varepsilon;) = m. \tag{2}$$

It is shown that for quantitative assessment of nature of dissipative processes and practical application, it is convenient to use effectiveness ratio of dissipation of energy (η):

$$\eta = 2m/(m + 1). \tag{3}$$

The coefficient η characterizes ability of structure of material to dissipate the brought mechanical energy in the course of hot deformation.

Size η changes in the range from zero to unit and is interpreted as the relative speed of production of entropy.

In the present article, results of the research characteristics of dissipation of energy in industrial alloys in the course of uniaxial stretching and compression on the example of near-alpha titanium alloy are stated.

During the planning and implementation, the present article used system approach which included the detailed analysis of structure of alloy before deformation and comparison of results of structural researches to results of mechanical tests and calculation of coefficient of dissipation of energy.

2. Materials and experimental methods

Mechanical tests of samples of titanium alloy cut from hot-rolled sheet products with thickness of 40 mm, the chemical composition of which is given in the **Table 1**, were made at different temperatures and speed parameters.

The initial microstructure of titanium alloy corresponded to a two-phase state which was created in the course of hot rolling.

The received structure is characterized by the large initial size of grains of a β-phase (~300 μm) and represents mix of the α-plates divided by β-phase layers (**Figure 1**).

Mechanical tests on stretching at room temperature were carried out on the tensile testing machine UEN30 "Shimadzu." At increased temperatures, the modernized universal testing machine UM5 was used.

Element content, % wt.									
Al	V	Mo	Fe	Si	C	O	H	N	Ti
5.4	2.0	1.2	0.25	0.3	0.1	0.15	0.08	0.04	The rest

Table 1.
Chemical composition of the studied alloy.

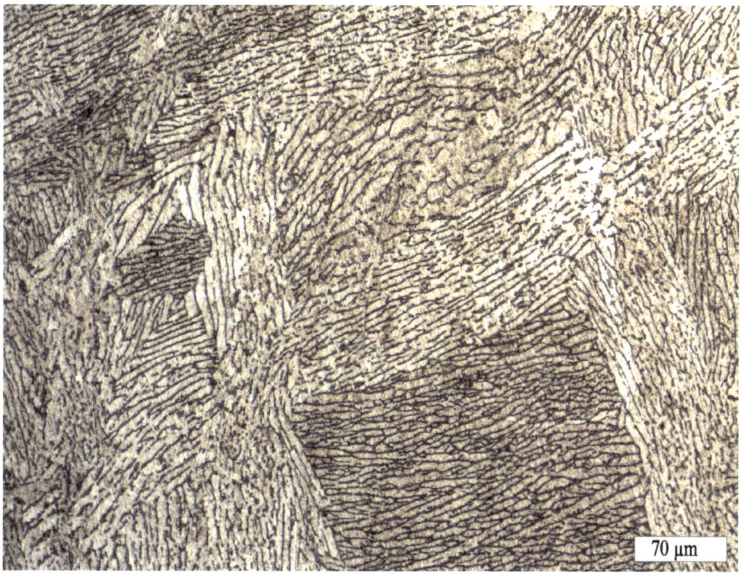

Figure 1.
Initial microstructure of alloy.

In an experiment, standard explosive samples with a diameter of 6 mm were used. For tests for compression cylindrical samples with a diameter of 5 and 10 mm on the high-temperature dilatometer DIL 805 were used.

At the same time deformation equaled ε = 0.3, and temperature of heating corresponded 800–1040°C. Average strain rate ε; was 10^{-3}–$10\ s^{-1}$.

Calculation of effectiveness ratio of dissipation of energy at deformation of samples under various temperature and high-speed conditions consisted in formation of a matrix of values of true tension at the set extent of deformation and their logarithms.

Calculation of coefficients of m and η and calculation of intermediate values of effectiveness ratio of dissipation of energy were made by a method of spline interpolation.

Results of calculation can be presented in the tabular, analytical, or graphic style.

The most evident is representation of results of calculation η in the form of 3D plot and cards of constant levels of effectiveness ratio of dissipation of energy. Calculation and creation of cards were carried out with the use of the Mathcad 15 program.

Microstructural researches were carried out on the polished samples of the deformed samples, which are cut out in the cross-sectional and longitudinal direction with application of modern methods [5, 6].

To get images, information technologies and specialized programs have been used ("Expert Pro", "Fractal") [7, 8].

3. Results of researches and discussion

According to results of mechanical tests, dependences are constructed $\sigma = f(\varepsilon)$.

The received dependences and their look do not contradict the settled ideas of behavior of metal polycrystals in the conditions of hot plastic deformation. So, in the course of process of plastic deformation of metal, tension smoothly increases and reaches a certain maximum (saturation) in which value is defined at the same time proceeding competing processes—hardenings and a weakening [9]. The growth rate of tension depends on temperature of heating and speed of deformation. At low temperatures and high speeds of deformation, flow stress continuously increases with deformation growth that is caused by the prevailing process of deformation hardening.

At the increased temperatures and low speeds of deformation, flow stress reaches a maximum and then goes down, reaching a certain constant value. In such type of charts, tension deformation is characteristic of the majority of the metals and alloys deformed at temperatures exceeding half the temperature of melting [10, 11].

Current tension size (σs) of the studied alloy depending on temperature and the speed of deformation is presented in **Table 2**.

Values of tension of a current at the set temperature and high-speed parameters of deformation were used for the subsequent calculation of effectiveness ratio of dissipation of energy η.

Change of coefficient η from temperature and high-speed parameters of deformation are presented in the form of the volume chart (**Figure 2**) and also in the form of the card of constant levels of effectiveness ratio of dissipation of energy (**Figure 3**).

Analyzing the results of change of effectiveness ratio of dissipation of energy presented on 3D plot and the map of constant levels of effectiveness ratio of dissipation depending on temperature and high-speed parameters of deformation, it is possible to note:

Temperature dependence $\eta = f(t, \varepsilon;)$: It is characterized by a maximum at temperatures 900–940°C. And with increase in speed of deformation, maximum shift toward big speeds of deformation is observed.

- The studied alloy is characterized by high efficiency of dissipation of energy in the studied range of temperature and high-speed parameters of process of hot deformation. Efficiency of energy of dissipation significantly does not change with increase in extent of deformation from 0.1 to 0.3.

- Mechanical properties of alloy at hot plastic deformation substantially depend on initial structure and temperature and high-speed parameters of deformation.

- In an initial state the studied alloy has the coarse-grained (not recrystallized) structure and the increased maintenance of a β-phase in comparison with an equilibrium state.

T, °C	σs, MPa			
	10^{-4} s^{-1}	10^{-3} s^{-1}	10^{-2} s^{-1}	10^{-1} s^{-1}
800	60	100	160	240
840	44	75	130	195
880	24	53	94	149
920	14	28	60	95
960	8.0	16	30	60
1000	5.8	10	18	32

Table 2.
The flow stress of examined alloy at various temperatures and strain rate values for tensile strain of $\varepsilon = 0.2$.

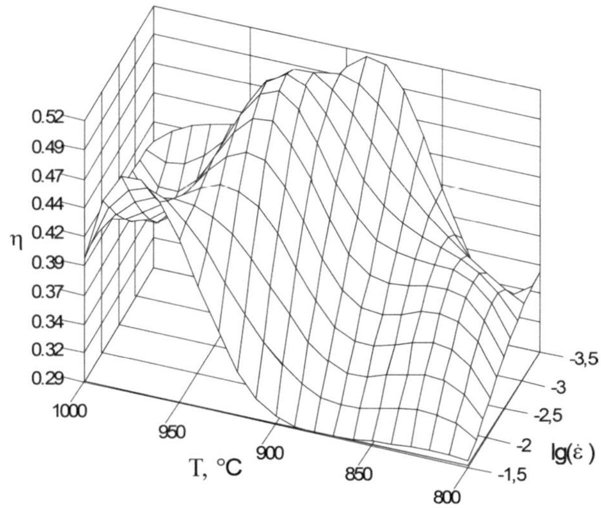

Figure 2.
3D plot of effectiveness ratio dissipations of energy at ε = 0.2.

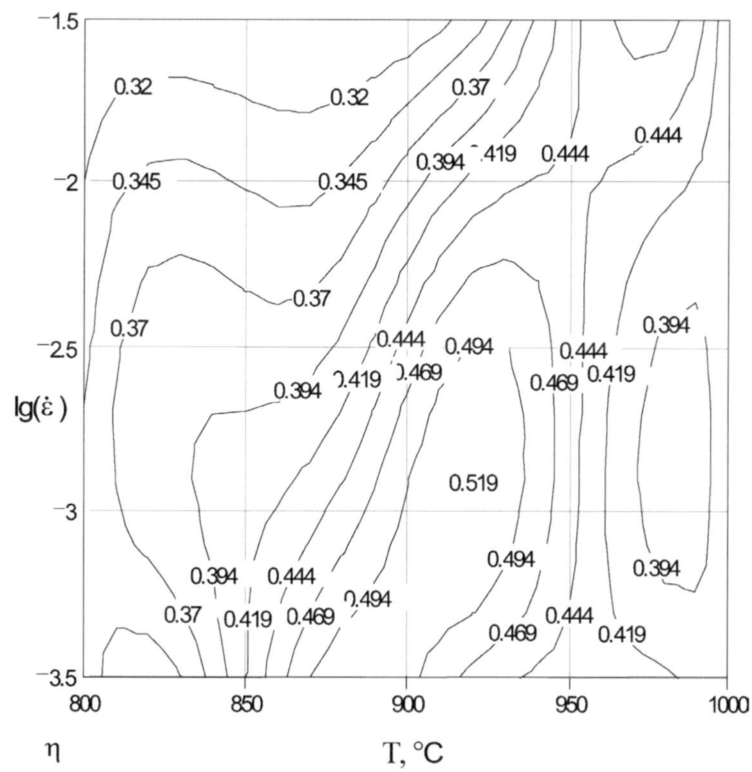

Figure 3.
Processing map for titanium alloy. Constant levels of coefficient efficiency of dissipation at ε = 0.2.

When heating alloy to temperature of 800°C, the first signs of recrystallization are observed, and further heating to temperatures of 920–940°C and endurance of 15 min. Process of recrystallization proceeds completely.

To process recrystallization, $\alpha \rightarrow \beta$ phase transformation is followed. An increase of the β-phase contents in the alloy when heated is represented in **Table 3**.

T, °C	750	800	850	900	950	1000	1040
β-phase, %	15	20	28	40	65	82	100

Table 3.
Change of volume fraction of a β-phase when heating alloy.

Increasing the heating temperature of the alloy leads to an increase in the phase change rate due to an increase in self-diffusion. The largest speed of phase transformation is observed at a temperature of heating of 900–950°C. The quantity of α- and β-phases decreases with temperature increase and increase in hold time. The quantity of a phase decreases with temperature increase and increase in hold time. At alloy heating temperatures (1040°C), the phase transformation which is followed by sharp integration of grains of a β-phase completely comes to the end (**Table 3**).

It follows from the provided data that the microstructure and phase composition of alloys undergo significant changes at a temperature of heating to temperatures more than 950°C owing to full completion of process of recrystallization and phase $\alpha \rightarrow \beta$ transformations.

Results of researches on the change of structure of alloy in the course of deformation at various temperatures and extent of deformation 0.4 are presented in **Table 4**.

Grain size change in phase $\alpha \rightarrow \beta$ transformation is characteristic for structure at hot deformation of alloy. If the size of grains α-phases decreases, then the size of grains β-phases on the contrary increases.

As appears from the data provided in **Table 4** in the course of deformation of alloy, there is an increase in quantity of β-phases. The considerable difference in the number of phases of composition of alloy is observed at deformation speeds 10^{-3} s^{-1}. Further increase in speed of deformation practically does not lead to significant change of quantity of β-phases in comparison with an initial condition of alloy. Increased rate of deformation results in intensive phase transformation in alloy at small degrees of deformation.

So, at extent of deformation $\varepsilon = 0.4$, quantity of β-phases reaches 12–20%, and at extent of deformation from 0.4 to 1.0, it reaches only 3–4%.

Change of phase composition in the course of deformation is often connected with intensity of diffusive processes. The authors of [11] note that heating to the temperature of deformation of the titanium alloy does not lead to the achievement of phase equilibrium. The reason for this phenomenon is the relatively low diffusion mobility of β-stabilizing elements. For example, the β-phase content is 49% at a strain temperature of 950°C and 30 minutes.

T, °C	Grain size, μm								Phase composition, %			
	$\varepsilon = 0$				$\varepsilon = 0.4$				$\varepsilon = 0$		$\varepsilon = 0.4$	
	I		II		I		II					
	α	β	α	β	α	β	α	β	α	β	α	β
840	12.7	4.7	6.8	2.4	6.6	5.5	3.6	2.5	70	30	54	46
900	11.0	4.9	6.6	3.0	7.9	7.7	4.3	4.2	60	40	42	58
960	10.4	6.7	6.8	5.1	9.9	12.5	6.7	6.9	35	65	22	78

Note: I, longitudinal section of a sample; II, the cross-section of a sample.

Table 4.
The size of grain and phase composition of alloy before deformation at various temperatures.

Practically the same quantity of β-phases is observed at 2.5 min. Endurance with extent of deformation $\varepsilon = 0.5$. This circumstance allows to make the assumption that not only the increase in diffusive mobility of atoms is caused by deformation but also temperature change of phase balance is the reason of phase transformation at action of external tension.

The phase $\alpha \leftrightarrow \beta$ transformation is accompanied by a volumetric effect. It is known that various authors estimate this value to be about 0.15% [12]. Transformation of $\alpha \leftrightarrow \beta$ is accompanied by a volumetric effect and $\alpha \rightarrow \beta$ transformation-negative volumetric effect. Therefore with an external pressure, there is a temperature change of polymorphic transformation. The speed of phase transformations generally depends on the difference of free energy of an initial and final state and also the size of change of volume upon this transition. As the size of free energy and volume depend on pressure, it is possible to expect that the speed of phase transformations will also depend on pressure.

In that case when phase transformations are carried out in the diffusive way, the kinetics of phase transformations is defined by change of speed of the course of diffusive processes with a pressure.

The driving force of phase $\beta \rightarrow \alpha$ transformation in titanium alloys is shift of phase equilibrium temperature under action of external tensions. The rate of phase change is determined by the diffusion mobility of the β stabilizing elements' atoms. The interesting fact established when studying changes of a microstructure of alloys at hot deformation is transformation of initial lamellar structure in granular, which is most brightly shown at a temperature of deformation of 920°C and strain rate of $1.1 \cdot 10^{-3}$ s^{-1} (**Figure 4**).

Grain shape coefficient was determined by quantitative

metallography method $K_{ф} = l_{\alpha}/d_{\alpha}$; where l_{α} is the length of the plates and d_{α} is the width of the plates α-phases. The results of the calculations showed that intensive change of grain shape occurs up to deformation of 100%; at higher deformation, Kf stabilizes at values of $\approx$ 1.2–1.5.

Equiaxial grains of structure 200 ÷ 300 microns in size are observed in the field of temperatures of β-phases ($t \geq 1040$°C).

The nature of dissipative processes described above finally defines indicators of plasticity of alloy. Maximum stability of plastic deformation of alloy is observed at compliance of temperature-speed deformation parameters and maximum

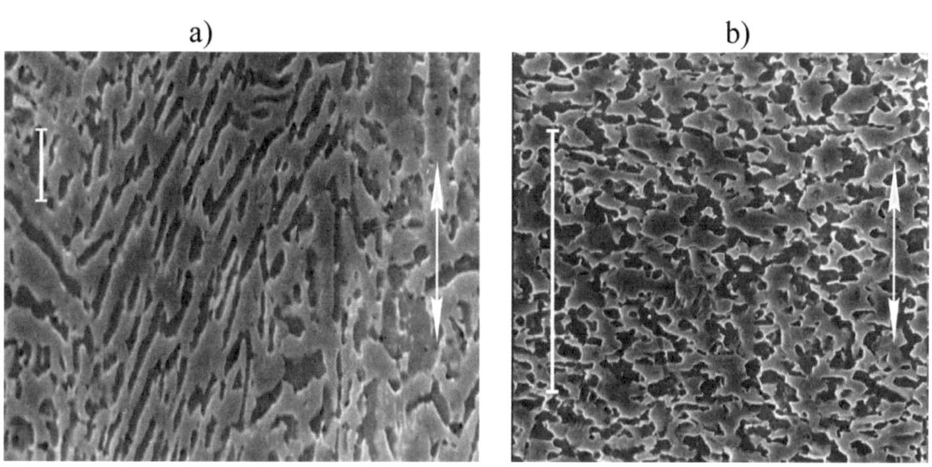

Figure 4.
An alloy microstructure after deformation at 920°C and speeds of deformation $1{,}1{\cdot}10^{-3}$ s^{-1}. (a) δ = 55%, (b) δ = 200%. ×1000.

coefficient of energy dissipation efficiency [10]. All signs of superplasticity state are observed at temperature of 920-960°C and deformation rate of 10^{-3}–10^{-2} s^{-1}.

Thus, on the basis of the analysis of structural changes when heating and plastic deformation of alloy, it is possible to draw the following conclusions.

4. Conclusions

1. At hot plastic deformation of the studied titanium alloy, there are, at least, two dissipative processes—dynamic recrystallization and phase transformation.

2. The maximum of efficiency of dissipation corresponds to the simultaneous balanced course of these two processes.

3. The temperature and high-speed extremum of effectiveness ratio of a dissipation of energy corresponds to conditions of the maximum stability at a plastic strain of alloy.

4. Results of researches about development and speed of dissipative processes at a hot plastic strain of the studied alloy can be used for optimization of the technological modes of hot treatment by pressure.

Author details

Mikhail Mikhaylovich Radkevich[1*], Nikolay Rafailovich Vargasov[1] and Boris Konstantinovich Barakhtin[2]

1 Peter the Great Saint Petersburg Polytechnic University, Russia

2 Saint Petersburg State Marine Technical University, Russia

*Address all correspondence to: radmich@mail.ru

References

[1] Kaibyshev OA. Superplasticity of Industrial Alloys. Moscow: Metallurgy; 1984. 284 p. (in Russian)

[2] Poirier JP. High Temperature Plasticity of Metallic Bodies. TRANS. FR. Moscow: Metallurgy; 1982. 273 p. (in Russian)

[3] Prasad YVRK, Sasidhara S, editors. Hot Working Guide A Compendium of Processing Maps. Bangalore: Department of Metallurgy, Indian Institute of Science; 2004. 560 p

[4] Vargasov NR, Rybin VV. Optimization of temperature-rate modes of plastic deformation on the criterion of dissipation of mechanical energy. Metallography and Heat Treatment of Metals. 1999;**9**:52-56. (in Russian)

[5] Vargasov NR, Rybin VV. Accumulation and dissipation of energy by hot plastic deformation of titanium alloy. Materials Science. 1999;**1**(18):63-69. (in Russian)

[6] Barakhtin BK, Nemets AM. Metals and alloys. Analysis and research. Physicoanalytical methods of study of metals and alloys. Nonmetallic Inclusions NPO "Professional", St. Petersburg; 2006. 490 p. (in Russian)

[7] Barahtin BK, Chashnikov VF. A computer program for multifractal analysis of images of structures of metals and alloys. Problems of Materials Science. 2001;**N4**(28):5-8. (in Russian)

[8] Barahtin BK, Vargasov NR. New approaches in assessing the structural and mechanical condition of the low alloy rail steel during thermal treatment. Materials Science and Engineering. 2014;**12**:8-14. (in Russian)

[9] Radkevich MM. Technology Strengthening Programme Mechanics and Heat Treatment. St. Petersburg: Technical Publishing, St. Petersburg State Technical University; 2011. 263 p. (in Russian)

[10] Vargasov NR, Barahtin BK. New approach to the assessment of the structural condition of rail steel during hot deformation. In: Materials of 4th International Scientific-Practical Conference. The Modern Mechanical Engineering. Science and Education. Peter the Great, St. Petersburg Polytechnic University. 2014. pp. 1095-1104. (in Russian)

[11] Radkevich MM. Peculiarities of structure and mechanical properties of soft mechanical and thermal processing of scientific and technical statements of St. Petersburg. Science and Education, St. Petersburg; 2008. Peter the Great, St.Petersburg Polytechnic University No. 1; 2012. p. 142 (in Russian)

[12] Titanium alloys. Metallography of titanium alloys. Moscow: Metallurgy; 1980. 464 p. (in Russian)

Chapter 6

Surface Treatment of Titanium Alloys in Oxygen-Containing Gaseous Medium

Vasyl Trush, Viktor Fedirko and Alexander Luk'yanenko

Abstract

The aim of investigations on the chapter was to determine regularities of solid solution hardening of surface layers of titanium alloys depending on the conditions of thermodiffusion saturation in rarified gas medium containing oxygen and determine the correlations between parameters of surface-hardened layers (surface hardness, depth of hardened zone, microstructure) and fatigue properties of titanium alloys under various methods of surface hardening. To achieve the formulated aim, the following methods were used: (a) thermodiffusion saturation of titanium alloys in rarified gas medium containing oxygen in the wide range of temperature-time and gas-dynamical parameters and (b) surface deformation by ultrasonic shock and shot-blasting treatments with rapid annealing of deformed surface by means of induction heating. The positive influence of surface hardening on the fatigue characteristics is decreased under the increasing of l when K is constant. The highest relative gain of fatigue strength ($\Delta\sigma_{-1}$) of samples with CTT surface-hardened layers is marked for the low- and middle-strong alloys VT1-0 and OT4-1. Thus for alloy VT1-0, $\Delta\sigma_{-1}$ = 35% under relative gain of surface hardness K = 70% and l = 3 0 μm. For the near-α-alloy OT4-1, $\Delta\sigma_{-1}$ = 38% under relative gain of surface hardness K = 35% and l = 4 5 ...50 μm.

Keywords: titanium alloy, oxygen, mass gain, solid-solution hardening, fatigue strength

1. Introduction

An increasing of fatigue strength and endurance of titanium alloys of various structural classes can be achieved by the formation of surface layers with regulated structure and phase state during surface hardening processes of chemico-thermal treatment (CTT). CTT in the present time the thermodiffusion hardening of surface layers by interstitial impurities is not used to increase the fatigue properties of workpieces made of titanium alloys. However, works performed at PMI NASU have shown that the range of the parameter ***K*** exists where the endurance of alloys with gas-saturated layers is higher than fatigue endurance in the initial state. An optimal level of surface hardening depends on the phase-structural state of metal and relative depth of gas-saturated zone. The problem arises in the controlling of intensity of the physical and chemical processes in a "titanium alloy/gas medium" system to obtain the optimal phase-structural state of surface layers and increase the operating characteristics of metal. Therefore the aim of investigations on the second stage of the project was to determine (a) regularities of solid solution hardening of surface layers of titanium alloys (α, near-α, α + β) depending on the conditions of thermodiffusion saturation in rarified gas medium containing oxygen and (b) general regularities of influence of methods and regimes of surface deformation on phase, structural, and substructural state of various titanium alloys. It is expected that obtained results will allow to forecast the influence of the regime of CTT on the phase-structural state of surface layers of metal and level of hardening and determine the parameters of thermal treatments for achieving the regulated level of hardening.

Increasing of fatigue strength and durability of titanium alloy workpieces remains an actual modern problem. It is known that fatigue properties of titanium alloys can be increased sufficiently by means of optimization of phase-structural state of surface layer. The aim of the investigations of paper was to determine the correlations between parameters of surface-hardened layers (surface hardness, depth of hardened zone, microstructure) and fatigue properties of titanium alloys VT1-0, VT5, OT4-1, VT16, and VT22 under various methods of surface hardening: thermodiffusion saturation in gas medium containing oxygen (CTT). Determination of such correlations allow to define the parameters of surface hardening of titanium alloys necessary for increasing of fatigue properties and approximate these methods of surface hardening to the practical application.

2. Regularities of solid solution hardening of surface layers of titanium alloys (α, near-α, and α + β) depending on conditions of thermodiffusion saturation in rarefied gaseous medium containing oxygen

Accordingly, with the results obtained previously, the required level of solid solution hardening of surface layers of titanium alloys by interstitial impurities depends on the phase-structural state of metal and relative depth of hardened (gas-saturated) zone. In this connection the problem of purposeful control of intensity of physicochemical processes in the system "titanium (titanium alloy)/gas medium" for the formation of required phase-structural state of surface layers aimed to ensure the corresponding operating performance of materials (fatigue strength, endurance, etc.) arises. In turn, this envisages the study of kinetic regularities of interaction of titanium alloys in rarefied gas medium.

2.1 Kinetic parameters of interaction of VT1-0, VT5, OT4-1, and VT16 alloys with rarefied gas medium

As a result of interaction of titanium alloys with rarefied gas medium containing oxygen, the processes of sublimation and phase formation may take place [1]. The processes of gas saturation and phase formation will be predominated at selected temperatures (923...1023 K) and pressures (6.6×10^{-3}...6.6×10^{-2} Pa), according to the analysis of changing of free energy of formation of solid solution of oxygen in titanium and titanium monoxide (**Figure 1**).

At the individual case, for alloys alloyed by manganese, molybdenum, and vanadium, the sublimation is possible due to the formation of oxide compounds with high pressure of saturated vapor [2]. Thus the gas saturation and phase formation increase the specimen mass of investigated titanium alloy, while sublimation decreases.

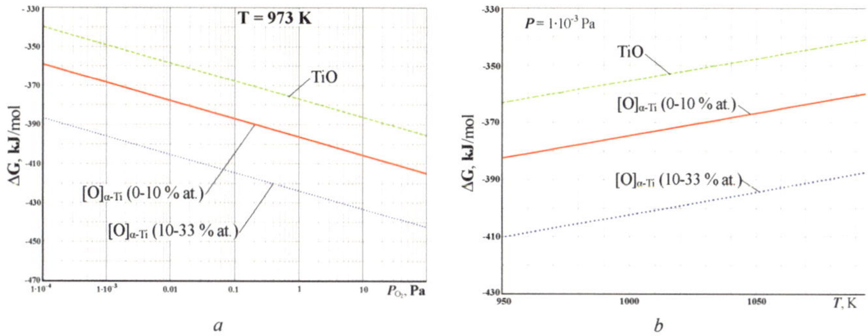

Figure 1.
Changes of free energy of formation of oxygen solid solutions in alpha-titanium and titanium monoxide as a function of (a) oxygen pressure and (b) temperature [1].

The influence of temperature-time and gas-dynamical parameters (***T*** = 650, 700, 750°C, $\boldsymbol{\tau}$ = 1, 3, 5 h, ***P*** = 6.6×10^{-3}, 1.33×10^{-2}, 6.6×10^{-2} Pa) of thermodiffusion saturation in controlled gas medium on the regularities of interaction of titanium alloys VT1-0, VT5, OT4-1, and VT16 is investigated by means of gravimetric analysis. The kinetic parameters of interaction of investigated titanium alloys determined by means of gravimetric analysis are shown in **Tables 1–4**.

According to the obtained results for α-titanium alloys (VT1-0, VT5), the process of gas saturation is intensified with the increasing of interaction temperature and pressure of gas medium (partial pressure of oxygen). The α-alloy VT1-0 (technical titanium) has the largest rate of interaction with rarefied gas medium containing oxygen under the all conditions. Alloying of titanium by 5% Al-alloy VT5-slows down slightly the rate of mass gain under the same conditions of interaction.

The mass loss caused by intensification of sublimation of alloying element Mn is possible for near-α-alloy OT4-1 (2% β-phase) under definite conditions of interaction with rarefied gas medium. The ratio of parameters ***T*** and ***P*** exists for this alloy when the rates of the gas saturation and sublimation processes become comparable. For the predomination of gas saturation processes, it is necessary to increase the partial pressure of oxygen or decrease the temperature of interaction.

Rate of gas saturation in rarefied gas medium decreased substantially with the increasing of quantity of β-phase in alloys (VT16 → VT22). This is caused by the decreasing of maximal solubility of oxygen in β-phase of titanium (6 at.%) in comparison with α-phase (33 at.%). With the increasing of interaction temperature, this difference appeared most appreciably. Therefore, it was concluded that alloys with large quantity of β-phase are the ones less sensitive to the changing of the conditions of interaction with rarefied gas medium containing oxygen.

***T*, °C**	**Δ*M*/*S* (μg/cm²) at residual pressure of gas medium**								
	P **= 6.6 × 10⁻³ Pa**			***P*** **= 1.33 × 10⁻² Pa**			***P*** **= 6.6 × 10⁻² Pa**		
	1 h	**3 h**	**5 h**	**1 h**	**3 h**	**5 h**	**1 h**	**3 h**	**5 h**
650	3.46	9.905	15.98	4.58	12.9	20.61	7.95	21.34	33.14
700	8.109	22.96	36.82	10.67	29.68	47.07	18.25	48.11	73.9
750	17.416	48.83	77.85	22.82	62.68	98.64	38.45	99.49	151.23

Table 1.
The specific mass gain of specimens of titanium alloy VT1-0 as a result of interaction with rarefied gas medium containing oxygen.

T, °C	$\Delta M/S$ (μg/cm^2) at residual pressure of gas medium								
	P = 6.6 × 10^{-3} Pa			P = 1.33 × 10^{-2} Pa			P = 6.6 × 10^{-2} Pa		
	1 h	3 h	5 h	1 h	3 h	5 h	1 h	3 h	5 h
650	2.24	6.53	10.65	2.57	7.44	12.10	3.31	9.48	15.31
700	5.62	16.2	26.29	6.43	18.42	29.77	8.24	23.31	37.36
750	12.83	36.64	59.07	14.64	41.49	66.60	18.68	52.14	82.86

Table 2.
The specific mass gain of specimens of titanium alloy VT5 as a result of interaction with rarefied gas medium containing oxygen.

T, °C	$\Delta M/S$ (μg/cm^2) at residual pressure of gas medium								
	P = 6.6 × 10^{-3} Pa			P = 1.33 × 10^{-2} Pa			P = 6.6 × 10^{-2} Pa		
	1 h	3 h	5 h	1 h	3 h	5 h	1 h	3 h	5
650	−0.10	−0.31	−0.52	3.18	9.55	16.0	7.25	22.00	36.75
700	−0.16	−0.49	−0.82	4.98	15.0	25.0	11.50	34.75	57.75
750	−0.24	−0.73	−1.23	7.47	22.0	37.0	17.25	52.00	86.50

Table 3.
The specific mass change of specimens of titanium alloy OT4-1 as a result of interaction with rarefied gas medium containing oxygen.

T, °C	$\Delta M/S$ (μg/cm^2) at residual pressure of gas medium								
	P = 6.6 × 10^{-3} Pa			P = 1.33 × 10^{-2} Pa			P = 6.6 × 10^{-2} Pa		
	1 h	3 h	5 h	1 h	3 h	5 h	1 h	3 h	5 h
650	2.00	6.00	9.66	2.50	7.80	13.00	3.50	10.60	17.80
700	2.32	7.00	11.60	2.80	8.80	14.20	4.10	12.60	22.50
750	2.80	8.50	13.29	3.20	9.75	16.60	5.30	16.70	27.40

Table 4.
The specific mass gain of specimens of titanium alloy VT16 as a result of interaction with rarefied gas medium containing oxygen.

Let us calculate the kinetic parameters as the function of temperature accordingly with the data of mass changing of specimens of titanium alloys in rarefied gas medium containing oxygen.

All kinetic dependences in the 5-h interval at the residual pressure P = 1.33 × 10^{-2} Pa follow linear dependence (1) satisfactorily. This indicates that surface reactions at the "metal-gas" interface are the controlling stage of the processes [3]:

$$\Delta M/S = \left(k^P \cdot \tau \pm A\right) \cdot 10^{-2}, \ \left[\text{g/m}^2\right] \tag{1}$$

where k^P is the coefficient of linear rate under constant pressure and A is the confidence interval with a possibility of 0.98.

The coefficient of linear rate under isobaric conditions of thermally activated process depends on the absolute temperature T accordingly with Arrhenius equations [1]:

$$k^{P}(T) = B \cdot \exp\left(-E_{\text{екс}}/RT\right) \pm C,\ \left[\text{g m}^{-2}\ \text{h}^{-1}\right] \quad (2)$$

where B is the constant depending on temperature, $E_{\text{екс}}$ is the total energy of process activation, and C is the confidence interval with a possibility of 0.98. Constants for Eqs. (1) and (2) and experimental activation energy of process are presented in **Table 5**.

All kinetic regularities follow linear dependence satisfactorily under isothermal conditions and various pressures accordingly with the data of thermogravimetry:

$$\Delta M/S = \left(k^{T} \cdot \tau \pm F\right) \cdot 10^{-2},\ \left[\text{g/m}^{2}\right] \quad (3)$$

where k^{T} is the coefficient of linear rate under constant temperature (700°C) and F is the confidence interval with a possibility of 0.98. Dependence of coefficient of linear rate under isothermal conditions on the residual pressure of medium is approximated satisfactorily by logarithmic dependence:

$$k^{T}(P) = \left[H + J \cdot \ln(P)\right] \pm K,\ \left[\text{g m}^{2}\ \text{h}^{-1}\right] \quad (4)$$

where H and J are the constants depending on pressure and K is the confidence interval with a possibility of 0.98.

Coefficients for Eqs. (3) and (4) are presented in **Table 6**.

2.2 Effect of temperature and time on surface metal hardness and hardening zone depth

Formation of interstitial solid solution in the metal during diffusion saturation of titanium alloys by gases in rarefied gas medium (mainly by oxygen) is bound up with strong distortion of crystallographic lattice (**Figure 2**) and as a result of this with essential increasing of hardness of metal. Therefore, the parameters of gas-saturated layers were determined by means of two methods: using microhardness

Alloy	T, °C	At T = 700°C				
		Using formula (1)		Using formula (2)		
		k^{P}, g m^{-2} h^{-1}	A, g m^{-2}	B, g m^{-2} h^{-1}	$E_{\text{екс}}$, J mol^{-1}	C, g m^{-2} h^{-1}
BT1-0	650	3.23	1.67	4×10^{7}	7447.6	4.66
	700	7.46	3.53			
	750	15.8	4.21			
BT5	650	2.44	0.59	1×10^{8}	8140.6	3.91
	700	6.02	2.96			
	750	13.49	4.10			
BT16	650	2.6	0.1	29.21	1350.8	1.07
	700	2.86	0.87			
	750	3.3	1.10			
OT4-1	650	3.2	0.2	16585.2	4148.6	3.34
	700	5.0	1.41			
	750	7.38	2.87			

Table 5.
Kinetic parameters of gas saturation of titanium alloys under isothermal conditions.

Alloy	$P \times 10^2$ Pa	At $T = 700°C$				
		Using formula (3)		Using formula (4)		
		k^T, g m^{-2} h^{-1}	F, g m^{-2}	H, g m^{-2} h^{-1}	J, g m^{-2} h^{-1}	K, g m^{-2} h^{-1}
BT1-0	0.66	7.46	1.25	24.357	3.3804	1.05
	1.33	9.57	1.53			
	6.6	15.2	2.24			
BT5	0.66	5.3	0.64	10.243	0.98	0.77
	1.33	6.02	0.90			
	6.6	7.57	1.68			
BT16	0.66	2.32	0.24	6.8862	0.9116	0.37
	1.33	2.86	0.74			
	6.6	4.41	1.05			
OT4-1	0.66	−0.16	0.063	25.189	4.8973	2.11
	1.33	5.00	0.41			
	6.6	11.56	0.77			

Table 6.
Kinetic parameters of gas saturation of titanium alloys under isothermal conditions.

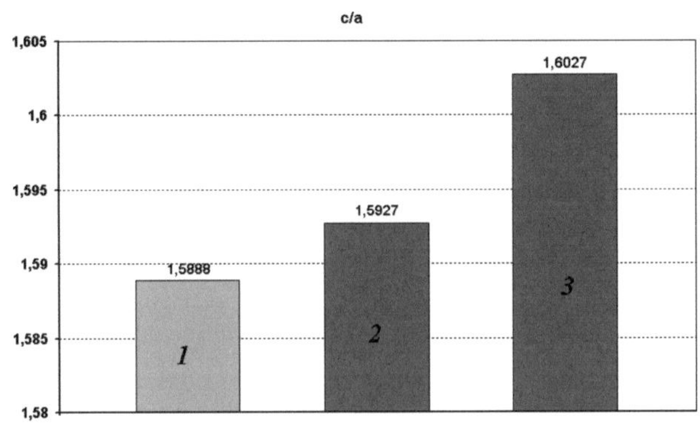

Figure 2.
*Changing of ratio of parameters of crystallographic lattice (**c/a**) of titanium alloy surface layer VT1-0 as a function of CTT regime: (1) in initial state, (2) after CTT: **T** = 650°C, **P** = 6.6 × 10^{-3} Pa, **τ** = 5 h, (3) after CTT: **T** = 750°C, **P** = 6.6 × 10^{-3} Pa, **τ** = 5 h.*

that was measured on both surface and cross section of metallographic sample made of the gas-saturated specimen and changing parameters of crystallographic lattice. The last one method is more complex and laborious; therefore, it was used only for determination of parameters of maximally hardened layer. Depth of gas-saturated layer recognizes as a distance from the surface where increasing of hardness caused by dissolution of oxygen is equal to the measurement error.

The experimentally obtained results of influence of parameters of thermodiffusion saturation on the gain of surface hardness K of investigated titanium alloys are presented in **Tables 7–14** ($K = ((H_\mu^s - H_\mu^c)/H_\mu^c) \times 100\%$, where H_μ^s is the surface hardness of metal and H_μ^c is the bulk hardness) and depth of gas-saturated zone l determined by durometry.

The results concerning the determination of parameters of crystallographic lattice of specimens of investigated alloys after different regimes of thermodiffusion saturation are presented in **Tables 15–18**.

T, °C	K (%) at residual pressure of gas medium								
	$P = 6.6 \times 10^{-3}$ Pa			$P = 1.33 \times 10^{-2}$ Pa			$P = 6.6 \times 10^{-2}$ Pa		
	1 h	3 h	5 h	1 h	3 h	5 h	1 h	3 h	5 h
650	41	66	81	56	85	103	80	121	143
700	37	61	75	49	78	96	74	113	134
750	22	37	47	30	48	60	45	72	88

Table 7.
Gain of surface hardness of titanium alloy VT1-0 as a result of interaction with rarefied gas medium containing oxygen.

T, °C	K (%) at residual pressure of gas medium								
	$P = 6.6 \times 10^{-3}$ Pa			$P = 1.33 \times 10^{-2}$ Pa			$P = 6.6 \times 10^{-2}$ Pa		
	1 h	3 h	5 h	1 h	3 h	5 h	1 h	3 h	5 h
650	30	47	58	37	58	70	52	78	92
700	28	45	56	35	56	67	50	75	89
750	6.5	11	15	10	18	22	20	32	40

Table 8.
Gain of surface hardness of titanium alloy VT5 as a result of interaction with rarefied gas medium containing oxygen.

T, °C	K (%) at residual pressure of gas medium								
	$P = 6.6 \times 10^{-3}$ Pa			$P = 1.33 \times 10^{-2}$ Pa			$P = 6.6 \times 10^{-2}$ Pa		
	1 h	3 h	5 h	1 h	3 h	5 h	1 h	3 h	5 h
650	38	66	86	33	58	75	11	18	24
700	17	30	38.9	28	50	64	26	45	59
750	9	17	22	8	15	19	13	23	29

Table 9.
Gain of surface hardness of titanium alloy OT4-1 as a result of interaction with rarefied gas medium containing oxygen.

T, °C	K (%) at residual pressure of gas medium								
	$P = 6.6 \times 10^{-3}$ Pa			$P = 1.33 \times 10^{-2}$ Pa			$P = 6.6 \times 10^{-2}$ Pa		
	1 h	3 h	5 h	1 h	3 h	5 h	1 h	3 h	5 h
650	25	47	60.2	24	41	54	22	38	48.8
700	5	9	11.8	3	5	6.7	1	2	2.5
750	0.45	0.77	1	0.4	0.8	0.76	0.4	0.8	1

Table 10.
Gain of surface hardness of titanium alloy VT16 as a result of interaction with rarefied gas medium containing oxygen.

The regularities of thermodiffusion saturation intrinsic for all investigated alloys were revealed basing on the analysis of obtained results, namely, parameters of gas-saturated layer H_μ^s, ΔH_μ^s, and l increase with the increasing of saturation time

T, °C	*l* (μm) at residual pressure of gas medium								
	$P = 6.6 \times 10^{-3}$ Pa			$P = 1.33 \times 10^{-2}$ Pa			$P = 6.6{\cdot}10^{-2}$ Pa		
	1 h	3 h	5 h	1 h	3 h	5 h	1 h	3 h	5 h
650	6	13	18	7	14	20	8	16	22
700	10	22	30	11	24	33	13	27	36
750	19	38	53	21	41	57	24	46	62

Table 11.
Dimension of gas-saturated layer on titanium alloy VT1-0 as a result of interaction with rarefied gas medium containing oxygen.

T, °C	*l* (μm) at residual pressure of gas medium								
	$P = 6.6 \times 10^{-3}$ Pa			$P = 1.33 \times 10^{-2}$ Pa			$P = 6.6 \times 10^{-2}$ Pa		
	1 h	3 h	5 h	1 h	3 h	5 h	1 h	3 h	5 h
650	5	11	16	6	12	18	7	14	20
700	7	17	25	9	20	28	11	22	31
750	8	22	32	12	28	40	17	35	49

Table 12.
Dimension of gas-saturated layer on titanium alloy VT5 as a result of interaction with rarefied gas medium containing oxygen.

T, °C	*l* (μm) at residual pressure of gas medium								
	$P = 6.6 \times 10^{-3}$ Pa			$P = 1.33 \times 10^{-2}$ Pa			$P = 6.6 \times 10^{-2}$ Pa		
	1 h	3 h	5 h	1 h	3 h	5 h	1 h	3 h	5 h
650	13	22	29	26	45	58	36	62	80
700	20	35	45	33	57	73	45	77	100
750	30	52	67	40	70	90	54	93	120

Table 13.
Dimension of gas-saturated layer on titanium alloy OT4-1 as a result of interaction with rarefied gas medium containing oxygen.

T, °C	*l* (μm) at residual pressure of gas medium								
	$P = 6.6 \times 10^{-3}$ Pa			$P = 1.33 \times 10^{-2}$ Pa			$P = 6.6 \times 10^{-2}$ Pa		
	1 h	3 h	5 h	1 h	3 h	5 h	1 h	3 h	5 h
650	19	48	71	22	53	76	26	60	85
700	3	10	33	5	40	70	27	73	109
750	2	5	40	1	55	100	37	108	163

Table 14.
Dimension of gas-saturated layer on titanium alloy VT16 as a result of interaction with rarefied gas medium containing oxygen.

under the same pressure of gas medium and temperature (**Figures 3** and **4**); the depth of gas-saturated zone ***l*** is increased, and value of relative gain of surface hardness ΔH_{μ}^{s} is decreased with the increasing of saturated temperature in the range 650–750°C (**Figure 5**).

Operating mode of CTT	*a*	*c*	*c/a*
Initial state	2.9481	4.6842	1.5889
750°C, 5.3 × 10^{-4} Pa, 5 h	2.9484	4.6860	1.5893
750°C, 1.3 × 10^{-2} Pa, 5 h	2.9496	4.6979	1.5927
750°C, 6.6 × 10^{-2} Pa, 5 h	2.948	4.7248	1.6027

Table 15.
Changing of parameter of crystallographic lattice of alloy VT1-0 as a result of interaction with rarefied gas medium containing oxygen.

Operating mode of CTT	*a*	*c*	*c/a*
Initial state	2.9286	4.6746	1.5962
750°C, 6,.6 × 10^{-3} Pa, 5 h	2.9281	4.6753	1.5967
750°C, 1.3 × 10^{-2} Pa, 5 h	2.9263	4.6729	1.5968
750°C, 6.6 × 10^{-2} Pa, 5 h	2.9329	4.7050	1.6042

Table 16.
Changing of parameter of crystallographic lattice of alloy VT5 as a result of interaction with rarefied gas medium containing oxygen.

Operating mode of CTT	*a*	*c*	*c/a*
Initial state	2.9427	4.6823	1.5911
750°C, 6.6 × 10^{-3} Pa, 5 h	2.9419	4.6856	1.5927
750°C, 1.3 × 10^{-2} Pa, 5 h	2.9426	4.6879	1.5931
750°C, 6.6 × 10^{-2} Pa, 5 h	2.9426	4.6911	1.5942

Table 17.
Changing of parameter of crystallographic lattice of alloy OT4-1 as a result of interaction with rarefied gas medium containing oxygen.

Operating mode of CTT	*a*	*c*	*c/a*	*Beta*
Initial state	2.9287	4.6674	1.5936	3.2265
750°C, 6.6 × 10^{-3} Pa, 5 h	2.9298	4.6707	1.5942	3.2266
750°C, 1.3 × 10^{-2} Pa, 5 h	2.9284	4.6687	1.5942	3.2273
750°C, 6.6 × 10^{-2} Pa, 5 h	2.9301	4.6706	1.5940	3.2280

Table 18.
Changing of parameter of crystallographic lattice of alloy VT16 as a result of interaction with rarefied gas medium containing oxygen.

The first regularity is connected with the increasing of concentration of interstitial impurity in surface layer of metal with time and its penetration on the greater depth. The second regularity can be explained by acceleration of withdrawal of interstitial impurities from the surface due to the increasing of its diffusivity in α- and β-titanium with the increasing of temperature (**Figure 6**). Under such conditions, the flow of oxygen from the medium to the metal surface becomes smaller in comparison with withdrawal flow owing to diffusion from the surface into the metal depth.

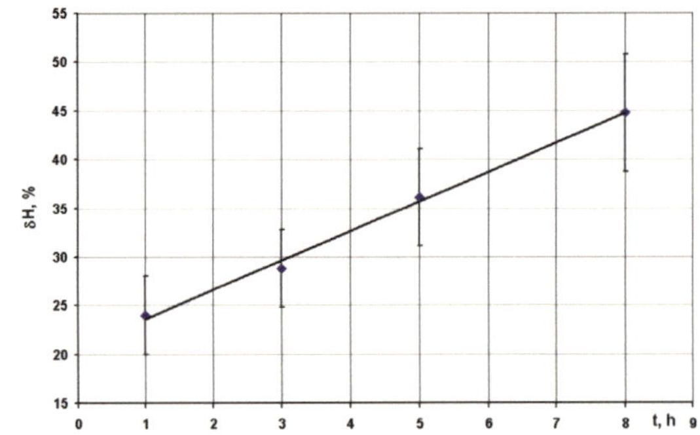

Figure 3.
Changing of the relative gain of surface hardness of titanium alloy VT1-0 under CTT ($\boldsymbol{T}$ = 750°C, $\boldsymbol{P}$ = 1.3 × 10^{-3} Pa) depending on exposure time.

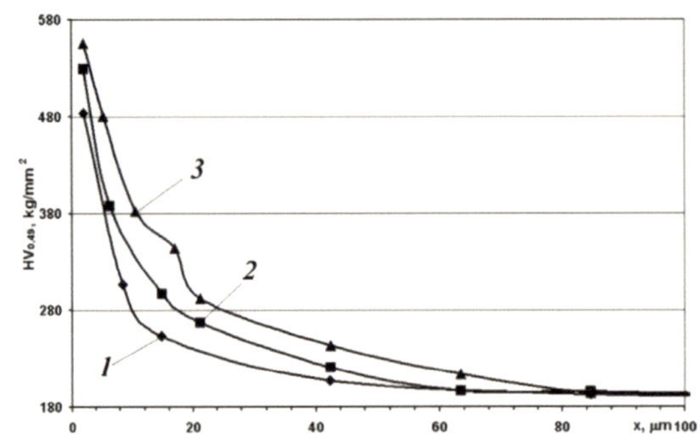

Figure 4.
Distribution of microhardness through the section of the specimens of alloy VT1-0 after CTT ($\boldsymbol{P}$ = 1.3 × 10^{-3} Pa, $\boldsymbol{\tau}$ = 5 h) at temperatures (1) 650°C, (2) 700°C, (3) 750°C.

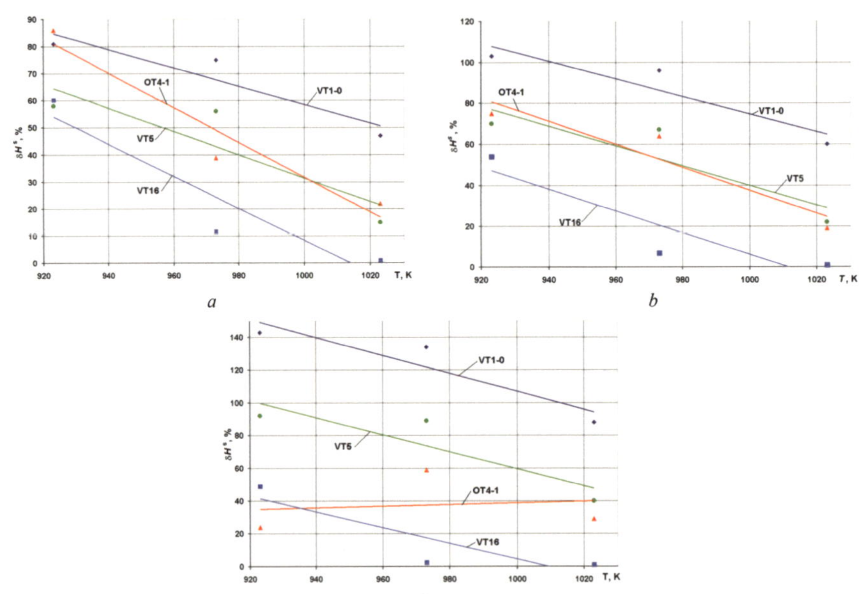

Figure 5.
Changing of the gain of surface hardness depending on interaction temperature with gas medium: (a) $\boldsymbol{P}$ = 6.6 × 10^{-3} Pa, (b) $\boldsymbol{P}$ = 1.3 × 10^{-2} Pa, (c) $\boldsymbol{P}$ = 6.6 × 10^{-2} Pa.

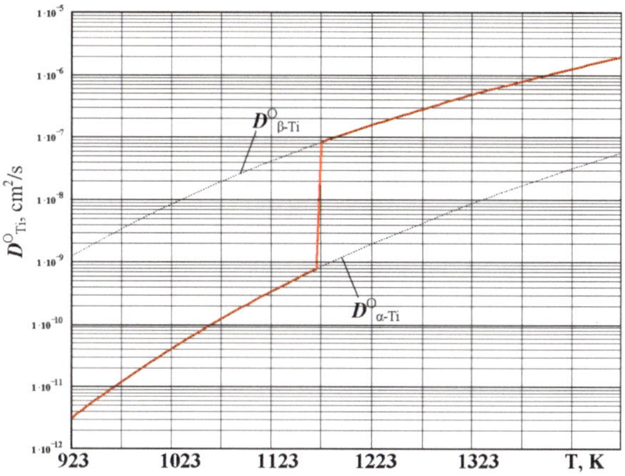

Figure 6.
Temperature dependence of oxygen diffusion coefficient in titanium [3].

The phase composition of titanium alloys influences essentially on the quantitative index of saturation at not influencing on the qualitative appearance of determined regularities. The maximal dissolution of oxygen in α-phase of titanium composes 33 at.%, while in β-phase, only 6 at.%. Therefore in the titanium alloys with a large content of α-phase (VT1-0, VT5, and near-α-alloy OT4-1) during thermodiffusion saturation in gas medium, the gas-saturated layers with high gradient of hardness are formed (**Figure** 7).

Gas-saturated layer with high gradient of hardness is formed on the (α + β)-alloy VT16 at the low temperatures of saturation (**Figure** 8).

The values of coefficient ***K*** representing the relative gain of surface hardness are higher sufficiently for α-alloys VT1-0 and VT5 in comparison with near-α-alloy OT4-1 and (α + β)-alloy VT16 under the same conditions of thermodiffusion saturation. On the other hand, the oxygen diffusion coefficient in β-titanium is high of order of magnitude in comparison with α-phase ($\boldsymbol{D}_{\beta\text{-Ti}} = 3 \times 10^{-10}$ cm²/c; $\boldsymbol{D}_{\alpha\text{-Ti}} = 2 \times 10^{-9}$ cm²/c at 800°C [3, 4]). Therefore, with the increasing in the alloys of volumetric content of β-phase, the depth of gas-saturated zone ***l*** is increased. Thus on the surface of alloy VT16, the gas-saturated layers with a depth 2–2.5 times bigger than α-alloy VT1-0 are formed (160 and 60 μm correspondingly after treatment under condition $\boldsymbol{T}$ = 750°C, $\boldsymbol{P} = 6.6 \times 10^{-2}$ Pa, $\boldsymbol{\tau}$ = 5 h).

Gas saturation and sublimation influence during thermodiffusion saturation not only the hardness but also the change of the state of surface and phase-structural state of near-surface layer of metal. Thus, because of sublimation and surface diffusion, the grain boundaries showed up; on some grains the characteristic step-like microrelief is developed (**Figure 9**).

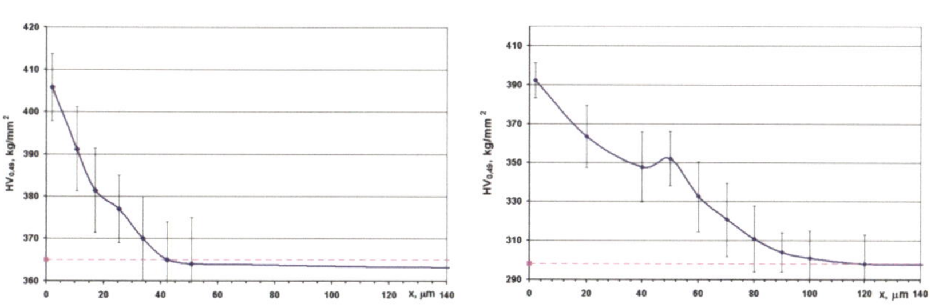

Figure 7.
*Distribution of microhardness in the surface layer of alloy (a) VT5 and (b) OT4-1 after CTT (**T** = 750°C, **P** = 6.6 × 10⁻² Pa, **τ** = 5 h).*

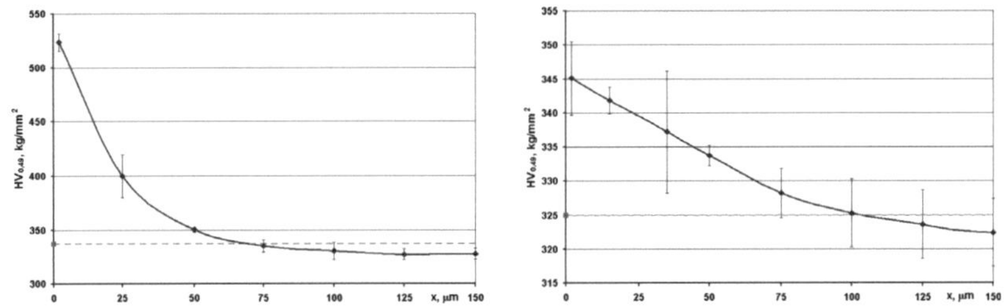

a *b*

Figure 8.
*Distribution of microhardness through the section of specimens of alloy VT16 after CTT (**P** = 1.3 × 10⁻² Pa, τ = 5 h) at temperatures (a) 650°C and (b) 750°C.*

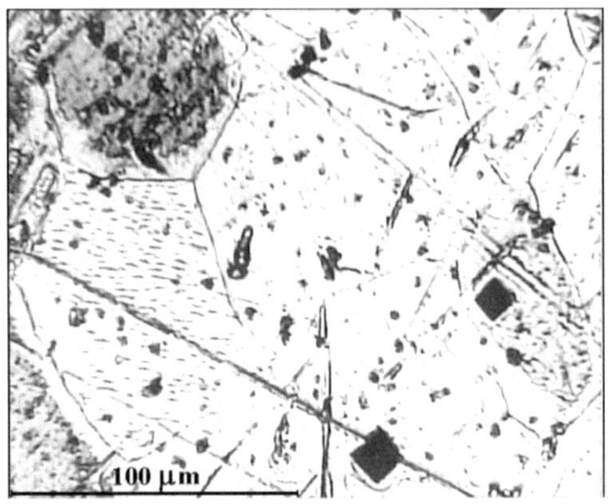

Figure 9.
*Microstructure of top surface of alloy VT1-0 after CTT (**T** = 700°C, **P** = 6.6 × 10⁻³ Pa, τ = 5 h). Larger indentation (loading 0.49 N) is obtained before thermal treatment, the smaller one after thermal treatment.*

Gas saturation stabilizes the α-phase of titanium in the surface layer of metal and increases its hardness. The diffusion layer consists of α-phase rich layer and transition layer. The α-phase rich layer differs in structure from the base metal by increased content of α-phase that is easily revealed with metallography (**Figure 10a**). This layer is represented often by only one α-phase. Transition layer is not visibly different from the base metal (**Figure 10b**), but for this layer the larger hardness is inherent.

Saturation of surface layer by oxygen and in some cases the sublimation of alloying elements lead to the redistribution of alloying elements in the surface layer

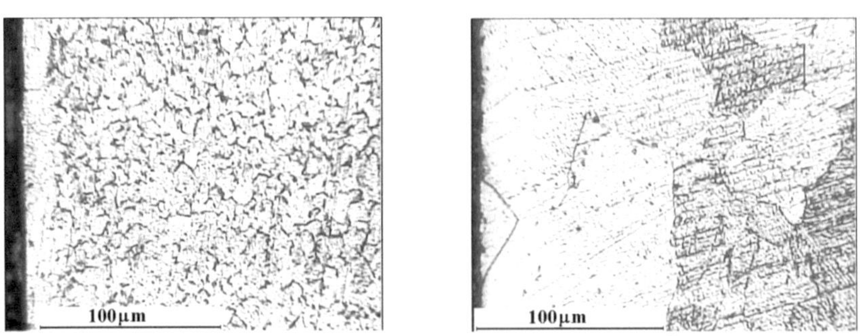

a *b*

Figure 10.
*Microstructure of surface layer of alloy OT4-1 (a) and VT1-0 (b) after CTT (**T** = 750°C, **P** = 6.6 × 10⁻³ Pa, τ = 5 h).*

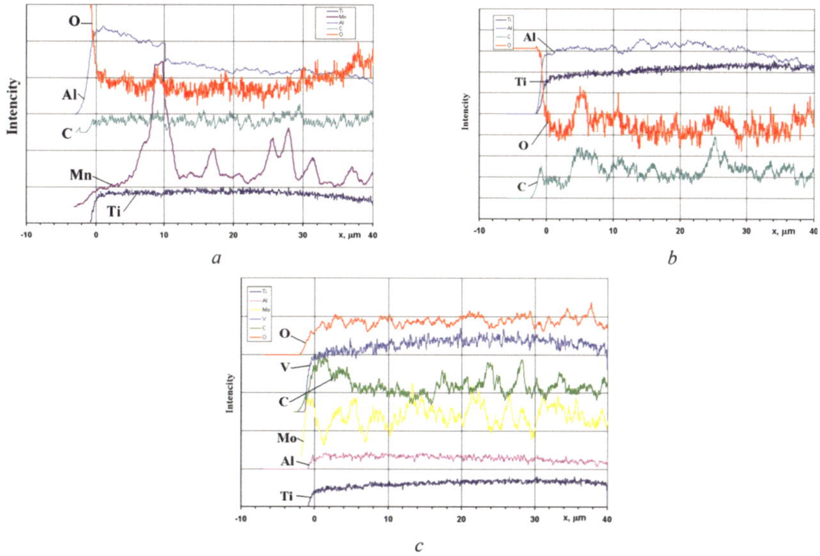

Figure 11.
Distribution of alloying elements in the surface layer of alloys (a) OT4-1, (b) VT5, and (c) VT16 after CTT (T = 750°C, P = 6.6 × 10^{-2} Pa, τ = 5 h).

of metal and changing of its chemical composition on the alloy OT4-1, but on the other alloys, such changes are not observed (**Figure 11**).

2.3 Effect of oxygen partial pressure on the surface hardness and hardening zone depth

Apart from the temperature-time parameters, the gas-dynamical parameters of gas medium (partial pressure of chemically active components and dynamics—I of leaking) influence the level of surface hardening. This influence should be taken into account during prediction of the consequences of thermodiffusion saturation of titanium alloys in the rarefied gas medium.

Let us consider in detail the influence of the changing of pressure of gas medium (P = 6.6 × 10^{-2}, 1.33 × 10^{-2}, 6.6 × 10^{-3} Pa, I = 5 × 10^{-5} Pa s^{-1}). The surface hardness and depth of hardened zone of most of titanium alloys are increased with the increasing of pressure under constant temperature accordingly with the obtained results (**Tables 13** and **14** and **Figure 12**). This is most appreciably for α-titanium alloys VT1-0 and VT5. The derivation from the regularity mentioned above is observed for near-α-alloy OT4-1 and (α + β)-alloy VT16 alloyed by elements with high volatility of oxides under saturation at 650°C (OT4-1, VT16) and 700°C (VT16)—decreasing of gain of surface hardness. In our opinion, this is caused by activation of sublimation process of Mn and V at relatively low intensity of gas saturation process and small solubility of oxygen in β-phase of titanium. At higher temperatures, the increasing of pressure of gas medium leads to the changing of tendencies in gain of surface hardness: the change in the inclination of temperature dependences is observed that can be connected with the increasing of oxygen flow from the medium, which becomes commensurable with the flow of withdrawal due to diffusion (**Figure 12c**).

Thermodiffusion saturation was performed under dynamic conditions of rarefied gas medium. That is to say, the residual pressure of medium is determined by dynamic equilibrium of gas flows pumped out and leaking into the reaction camera from the outside. The rate of leaking should be restricted because the increase of the

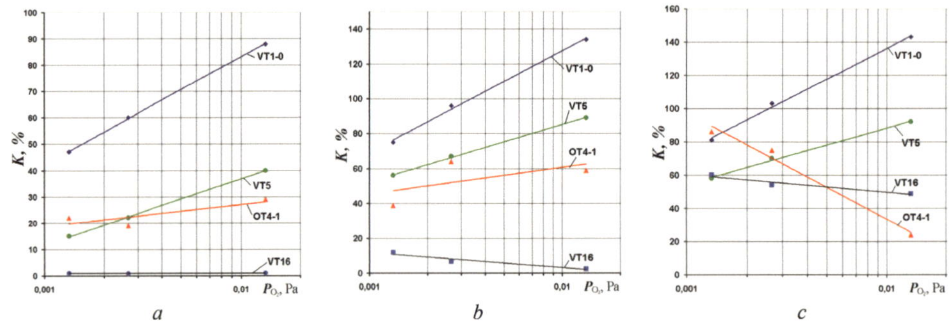

Figure 12.
The relative gain of surface hardness of titanium alloys for 5 h depending on the oxygen pressure in rarefied gas medium at temperatures (a) 650°C, (b) 700°C, and (c) 750°C.

flow of the leaking gases influences the kinetics of interaction similarly to the increasing of pressure [5]. It should be noticed that all mentioned results are obtained under the conditions of low enough specific rate of leaking into the reaction chamber of vacuum equipment—$\boldsymbol{I} = 5 \times 10^{-5}$ Pa s^{-1}. The increasing of leaking into the vacuum system intensifies sufficiently the oxidation and gas saturation of titanium alloys VT1-0, VT5, OT4-1, and VT16; however the regularities of this factor were not studied enough.

2.4 Relationship between the treatment parameters (pressure, temperature, duration) and level of interstitial solid

To predict the parameters of surface hardening of titanium alloys as a function of thermodiffusion saturation by interstitial impurities, the improved physico-mathematical model of gas saturation of titanium alloys in rarefied gas medium is proposed for using [6].

Intensity of thermodiffusion processes is determined by phase-boundary reaction and concentration distribution of diffusant and diffusion coefficient. Phase-boundary reaction consists of a number of processes occurring in vacuum: adsorption of molecules and atoms of gases of residual atmosphere, their dissolution in the metal, oxidation, etc.; rate of this reaction is changed depending on the degree of medium discharging. Actually, the boundary concentration of oxygen on the surface of the metal is not being steadied at once but is being increased gradually with the rate depending on the degree of gas medium discharging. Based on the thermodynamic analysis (see first milestone), the titanium oxides are in the equilibrium with pressure of oxygen at all degrees of discharging that provides the boundary solubility of oxygen in α-titanium, $\boldsymbol{C}_0$ = 33 at.% O_2, that causes formation of stoichiometric oxides [3, 4]. Thus the calculation of diffusion saturation of metal by oxygen must be performed basing on the next boundary condition [6]:

$$-D\frac{dC}{dx} = \alpha \cdot [C_0 - C(0,t)] \quad , \; x = 0 \tag{5}$$

where α is the coefficient of rate of phase-boundary reaction, $\boldsymbol{C}_0$ is the equilibrium concentration of oxygen in metal, $\boldsymbol{C}(0, \boldsymbol{t})$ is the actual concentration, and $\boldsymbol{t}$ is the time. These boundary conditions describe the mass change for period of time before the formation of oxide with certain thickness, which can already influence substantially the rate of the processes.

The function $\boldsymbol{C}(\boldsymbol{x}, \boldsymbol{t})$ is the known solution of the diffusion task:

$$C(x,t) = C_0 \cdot [\mathrm{erfc}(x/2\sqrt{Dt}) - \exp(hx + h^2 Dt) \cdot \mathrm{erfc}(x/2\sqrt{Dt} + h\sqrt{Dt})], \text{ where } h = \alpha/D \quad (6)$$

Mass changing of specimen under saturation by oxygen, dimension of diffusion zone up to 150 μm, and specimen thickness 3 mm per unit area is approximated by expression [5]:

$$M(t) = (C_0/h)[\exp(h^2 Dt)\mathrm{erfc}(h\sqrt{Dt}) - 1 + 2h\sqrt{Dt/\pi}]. \quad (7)$$

In the framework of physico-mathematical model of interaction of solid with gaseous medium under conditions of third type at the metal/gas interface, the coefficients of phase-boundary reaction (α) for alpha-titanium alloys VT1-0 and VT5 (see **Tables 19** and **20**) are determined using the experimental data (**Tables 1** and **2**).

The proposed approach for describing of the gas saturation processes with the use of coefficient of phase-boundary reaction α allows to calculate the mass change and concentration distribution of diffusant in the surface layer of metal under various temperature-time regimes of thermodiffusion saturation of α-titanium alloys, to determine the characteristics of gas-saturated layer (hardness distribution, depth of hardened zone) by using known correlation between hardness of surface layer and concentration of oxygen. The calculated nomograms for determination of permissible parameters of CTT of alloys VT1-0 and VT5 under condition of regulated level of surface hardening K = 25% which is presented in **Figure 13**.

The obtained analytical data are in a good accordance with the experimental results that allow using this approach for evaluation and prediction of parameters of thermodiffusion hardening of surface layer of α-titanium alloys.

The additional investigation for determination of model dependences for near-α and (α + β)-titanium alloys is necessary since at the selected temperatures the change in the ratio of phase components in alloys and sublimation of alloying elements during gas saturation are possible.

***T*, °C**	**α (cm/s) at residual pressure of gas medium**		
	***P* = 6.6 × 10^{-3} Pa**	***P* = 1.33 × 10^{-2} Pa**	***P* = 6.6 × 10^{-2} Pa**
650	5.017 × 10^{-9}	5.792 × 10^{-9}	7.565 × 10^{-9}
700	1.206 × 10^{-8}	1.393 × 10^{-8}	1.819 × 10^{-8}
750	2.662 × 10^{-8}	3.073 × 10^{-8}	4.014 × 10^{-8}

Table 19.
Coefficients of phase-boundary reaction (α) for titanium alloy VT1-0.

***T*, °C**	**α (cm/s) at residual pressure of gas medium**		
	***P* = 6.6 × 10^{-3} Pa**	***P* = 1.33 × 10^{-2} Pa**	***P* = 6.6 × 10^{-2} Pa**
650	1.76 × 10^{-9}	2.04 × 10^{-9}	2.66 × 10^{-9}
700	4.48 × 10^{-9}	5.18 × 10^{-9}	6.76 × 10^{-9}
750	1.04 × 10^{-8}	1.20 × 10^{-8}	1.56 × 10^{-8}

Table 20.
Coefficients of phase-boundary reaction (α) for titanium alloy VT5.

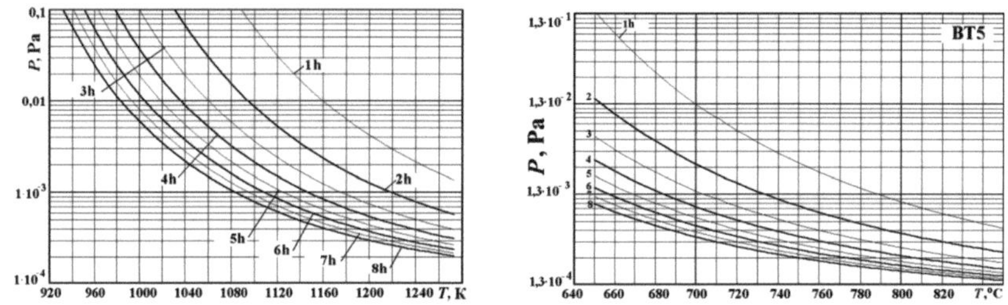

a b

Figure 13.
Nomograms for determination of parameters of CTT of titanium alloys (a) VT1-0 and (b) VT5 (the curves correspond to the level of surface hardening ***K*** *= 25%).*

3. Correlation between parameters of surface strengthening layers of different titanium alloys and their fatigue properties

3.1 Correlation between parameters of surface strengthening (surface hardness, hardening zone depth) and the fatigue properties of different titanium alloys under CTT

To determine the correlation between parameters of surface hardening and fatigue properties of titanium alloys of various structural classes under conditions of thermodiffusion saturation in controlled gas medium in the surface layers of metal, (a) the hardened layers of different depth with identical level of surface hardening, ***K*** = const, ***l*** = var., and (b) the hardened layers of identical depth with different level of surface hardening, ***K*** = var., ***l*** = const, were formed. It allows to reveal the influence of level and depth of surface hardening and also determine the optimal combination of these parameters.

3.1.1 α-Alloy VT1-0

3.1.1.1 Regulated surface hardening of VT1-0 titanium alloy under conditions of thermodiffusion saturation in controlled gaseous medium

To form the hardened layers with various ratios ***K*** and ***l***, the determined relations considering the influence of temperature-time and gas-dynamical parameters of rarefied gas medium containing oxygen on the parameters of hardened layers and peculiarities of solid solution surface hardening of titanium alloys of various structural classes were used [7–12]. The used approaches for selection of parameters of thermodiffusion saturation under specified ***K*** and ***l*** are demonstrated in **Figure 14** for alloy VT1-0 as an example.

Thus, the hardened layers of different depths with identical level of surface hardening can be obtained by means of changing the oxygen partial pressure and duration of saturation. The change of temperature influences the intensity of processes of thermodiffusion saturation. The influence of rate of oxygen leaking into the reaction chamber corresponds to the changing of partial pressure of chemically active component of gas medium (**Figure 15**). In addition the peculiarities of thermodiffusion saturation of titanium alloys of various structural classes should be taken into account. Thus, the hardened (gas saturated) layers with predetermined ratios ***K*** and ***l*** can be obtained due to changing of four parameters of thermodiffusion saturation ($\boldsymbol{T}$, $\boldsymbol{\tau}$, $\boldsymbol{P}_{O2}$, and $\boldsymbol{I}_l$—rate of oxygen in leakage into the reaction camera).

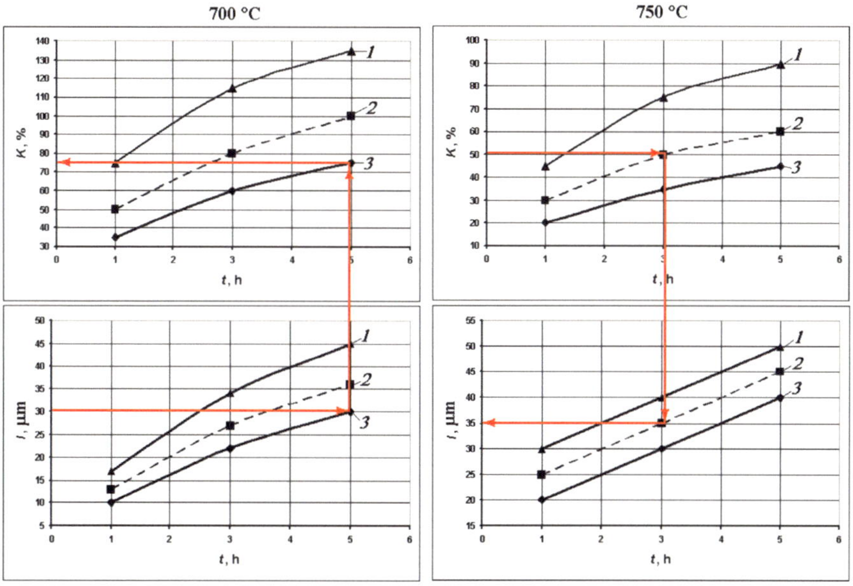

Figure 14.
Example of a choice of VT1-0 titanium alloy thermodiffusion saturation's parameters: (1) $\mathbf{P} = 6.6 \times 10^{-2}$ *Pa, (2)* $\mathbf{P} = 1.33 \times 10^{-2}$ *Pa, and (3)* $\mathbf{P} = 6.6 \times 10^{-3}$ *Pa.*

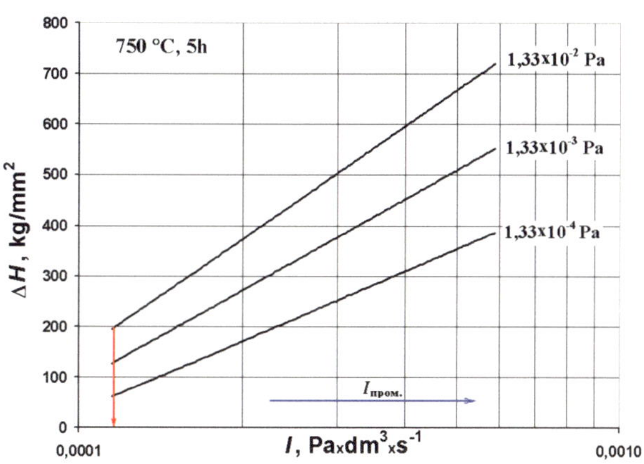

Figure 15.
Influence of inleakage rate ($\mathbf{I}_{in}$*) on VT1-0 titanium alloy surface hardness gain.*

The parameters of thermodiffusion saturation regimes of titanium alloy VT1-0 in controlled gas medium containing oxygen which are selected based on the approach mentioned above which are refined experimentally and correspond to normative documents [13] are presented in **Table 21**. Parameters of surface-hardened layer and hardness distribution through the cross section of samples are determined by means of durometry.

3.1.1.2 Influence of CTT on fatigue properties of α-titanium alloy VT1-0

The surface-hardened layers influence sufficiently the metal fatigue resistance. The results of fatigue tests of samples of alloy VT1-0 after regulated surface hardening by thermodiffusion saturation in gas medium are presented in **Figures 16–19** and **Tables 22** and **23**. The obtained results allow analyzing the influence of the level of surface hardening and depth of hardened zone on the metal fatigue resistance.

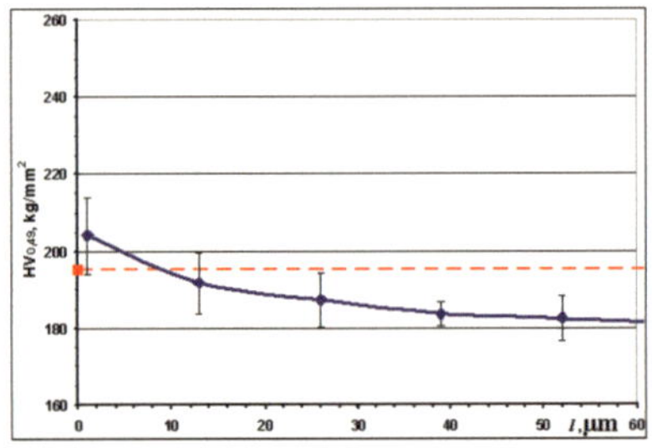

R0 – initial state: $\boldsymbol{K}$ = 5 %; $\boldsymbol{l}$ = 10 μm
annealing $\boldsymbol{T}$ = 800°C, $\boldsymbol{t}$ = 2 h, $\boldsymbol{P} = 5\times10^{-5}$ Pa, $\boldsymbol{I}_{in} = 0.1\cdot10^{-3}$ Pa·dm^3·s^{-1} + polishing +
annealing $\boldsymbol{T}$ = 600°C, $\boldsymbol{t}$ = 1 h, $\boldsymbol{P} = 5\times10^{-5}$ Pa, $\boldsymbol{I}_{in} = 0.12\times10^{-3}$ Pa·dm^3·s^{-1}

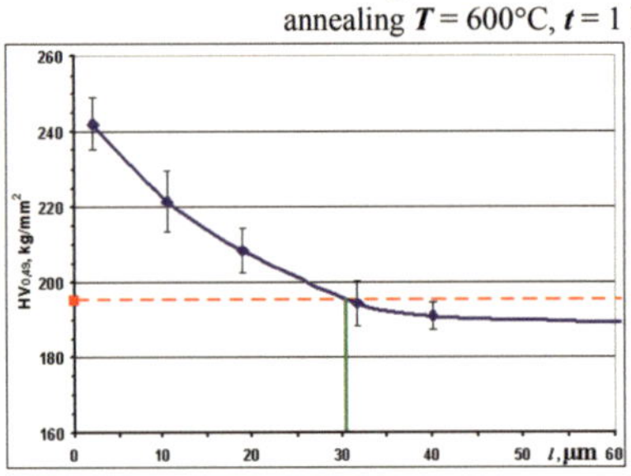

R1 - $\boldsymbol{K}$ = 25%; $\boldsymbol{l}$ = 30 μm
R0 + annealing $\boldsymbol{T}$ = 750°C, $\boldsymbol{t}$ = 3 h,
$\boldsymbol{P} = 6.6\times10^{-3}$ Pa, $\boldsymbol{I}_{in} = 0.12\cdot10^{-3}$ Pa·dm^3·s^{-1}

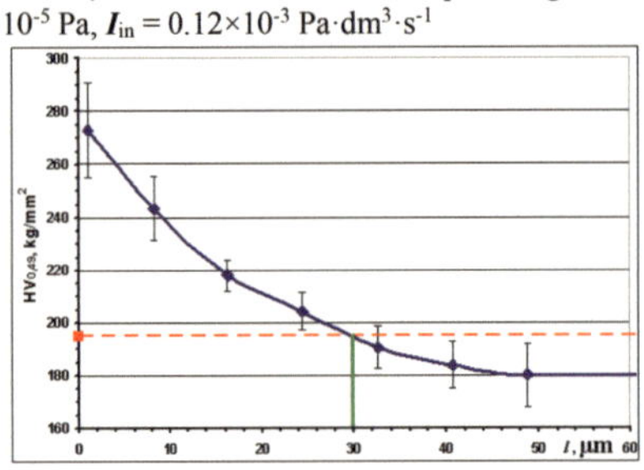

R2 - $\boldsymbol{K}$ = 50%; $\boldsymbol{l}$ = 30 μm
R0 + annealing $\boldsymbol{T}$ = 700°C, $\boldsymbol{t}$ = 5 h,
$\boldsymbol{P} = 6.6\times10^{-2}$ Pa, $\boldsymbol{I}_{in} = 0.12\cdot10^{-3}$ Pa·dm^3·s^{-1}

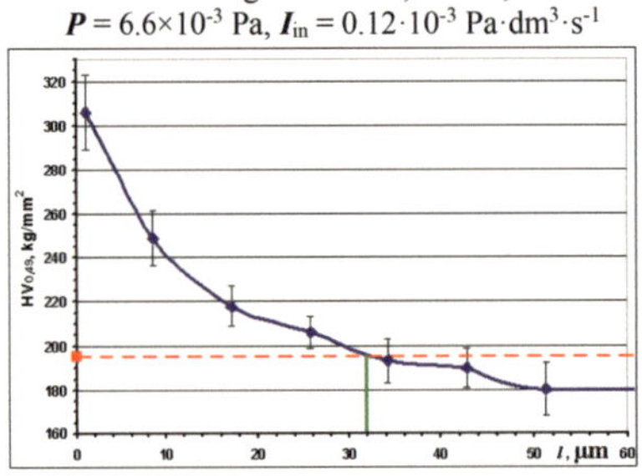

R3 - $\boldsymbol{K}$ = 70%; $\boldsymbol{l}$ = 30 μm
R0 + annealing $\boldsymbol{T}$ = 750°C, $\boldsymbol{t}$ = 3 h,
$\boldsymbol{P} = 6.6\times10^{-2}$ Pa, $\boldsymbol{I}_{in} = 0.12\cdot10^{-3}$ Pa·dm^3·s^{-1}

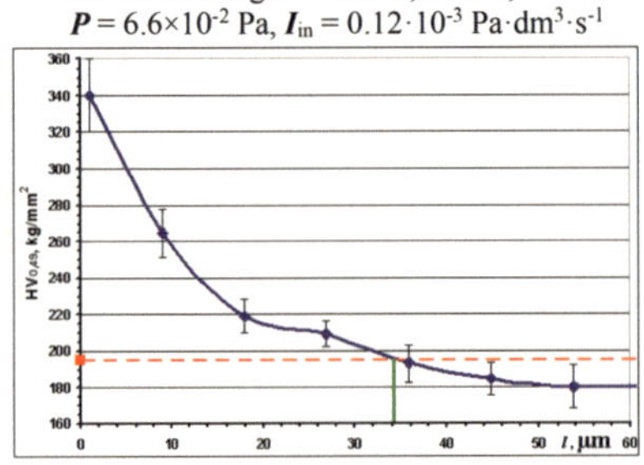

R4 - $\boldsymbol{K}$ = 90%; $\boldsymbol{l}$ = 30 μm
R0 + annealing $\boldsymbol{T}$ = 750°C, $\boldsymbol{t}$ = 3 h,
$\boldsymbol{P} = 6.6\times10^{-2}$ Pa, $\boldsymbol{I}_{in} = 0.16\cdot10^{-3}$ Pa·dm^3·s^{-1}

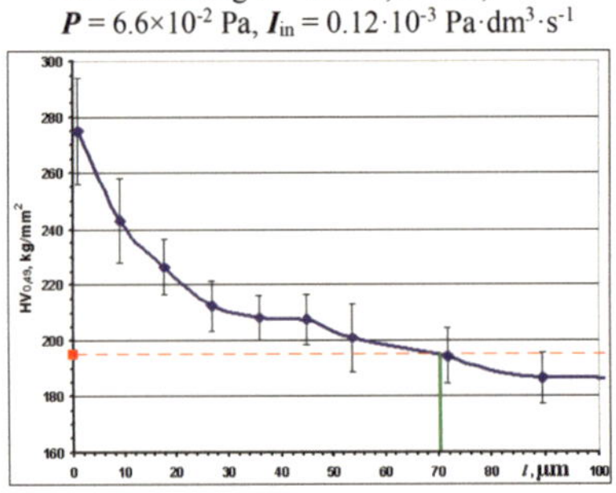

R5 - $\boldsymbol{K}$ = 50%; $\boldsymbol{l}$ = 70 μm
R0 + annealing $\boldsymbol{T}$ = 750°C, $\boldsymbol{t}$ = 8 h,
$\boldsymbol{P} = 3.9\times10^{-3}$ Pa, $\boldsymbol{I}_{in} = 0.12\cdot10^{-3}$ Pa·dm^3·s^{-1}

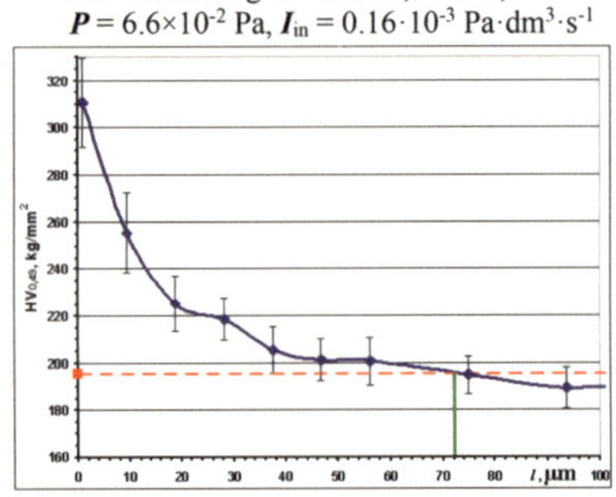

R6 - $\boldsymbol{K}$ = 70 %; $\boldsymbol{l}$ = 70 μm
R0 + annealing $\boldsymbol{T}$ = 750°C, $\boldsymbol{t}$ = 8 h,
$\boldsymbol{P} = 3.9\times10^{-3}$ Pa, $\boldsymbol{I}_{in} = 0.16\cdot10^{-3}$ Pa·dm^3·s^{-1}

Table 21.
Parameters of VT1-0 titanium alloy surface-hardened layers and thermodiffusion saturation's regimes.

3.1.1.3 Influence of level of surface hardening

Fatigue strength σ_{-1} of titanium alloy VT1-0 is being increased initially (**Table 2** and **Figure 17**) and then decreased with the increasing of level of surface hardening *K* from 5 to 90%, at constant depth of hardened zone ($\boldsymbol{l}$ = 30–35 μm). In other

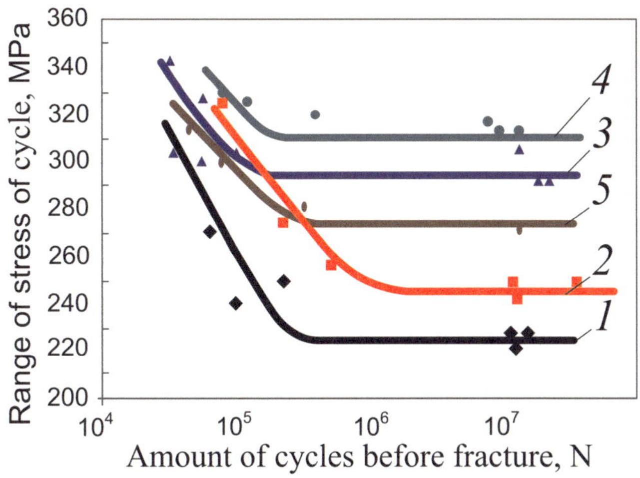

Figure 16.
*Fatigue curves of titanium alloy VT1-0, under rotating bending conditions, depending on the level of surface hardening when depth of hardened zone (gas saturated) is constant: (1) initial state **K** = 5%; **l** = 5 μm; (2) **K** = 25%; **l** = 30 μm (3) **K** = 50%; **l** = 30 μm (4) **K** = 70%; **l** = 30 μm (5) **K** = 90%; **l** = 30 μm.*

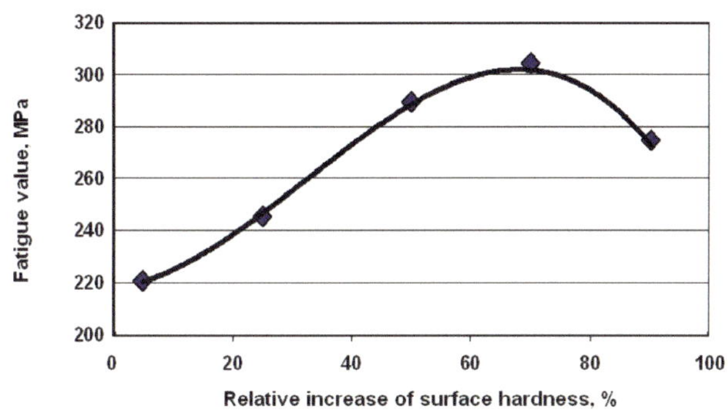

Figure 17.
*Fatigue strength of titanium alloy VT1-0, under rotating bending conditions, as a function of level of surface hardening when depth of hardened zone (gas saturated) is constant **l** = 30 μm.*

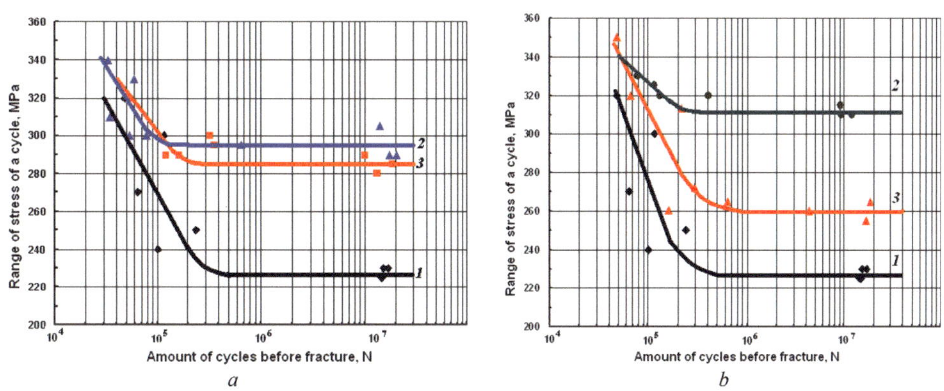

Figure 18.
*Fatigue curves of titanium alloy VT1-0, under rotating bending conditions, depending on depth of hardened (gas saturated) zone, when level of surface hardening is constant (a, **K** = 50%; b, **K** = 70%): (1) initial state **K** = 5%; **l** = 5 μm, (2) **l** = 30 μm, (3) **l** = 70 μm.*

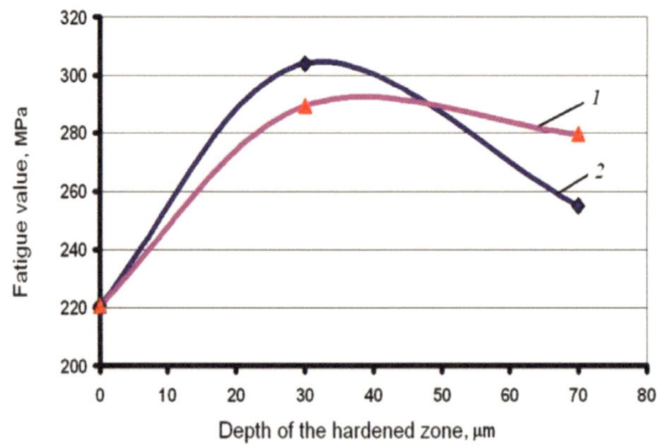

Figure 19.
*Fatigue strength of titanium alloy VT1-0, under rotating bending conditions, as a function of depth of hardened zone, when level of surface hardening **K** is constant: (1) **K** = 50% (2) **K** = 70%.*

#	Hardness of matrix H_O, MPa	Hardness of surface H_S, MPa	Average relative gain of surface hardness K, %	Depth of hardened zone l, μm	Fatigue strength σ_{-1}, MPa	Relative gain of fatigue strength $\Delta\sigma_{-1}$, %
R0	1800 (initial state)	1900	5	≤5	≤225	0
R1	1800	2250	25	30–35	≤245	8.9
R2		2700	50	30–35	≤295	31.1
R3		3050	70	30–35	≤310	37.7
R4		3420	90	30–35	≤275	22.2

Table 22.
*Fatigue strength of alloy VT1-0, under rotating bending conditions, depending on level of surface hardening **K**, when depth of hardened zone is **l** is constant.*

#	Hardness of matrix H_O, MPa	Hardness of surface H_S, MPa	Average relative gain of surface hardness K, %	Depth of hardened zone l, μm	Fatigue strength σ_{-1}, MPa	Relative gain of fatigue strength $\Delta\sigma_{-1}$, %
1.	1800 (initial state)	1900	5	5	≤225	0
2.	1800	2700	50	30	≤295	31.1
3.		2700	50	70	≤285	28.4
4.		3050	70	30	≤310	37.7
5.		3050	70	70	≤260	15.5

Table 23.
*Fatigue strength of titanium alloy VT1-0, under rotating bending conditions, depending on depth of hardened zone **l**, when level of surface hardening **K** is constant.*

words, fatigue strength has a maximum. Relative gain of fatigue strength $\Delta\sigma_{-1}$ is the highest when K = 70%. Such character of changing σ_{-1} can be explained by the improvement of fatigue properties due to dissolution of oxygen in metal with formation of solid solution that is accompanied by the appearing of compressing stress. On the other hand, metal is embrittled due to solid solution hardening by oxygen dissolution. One or another factor dominates the defined conditions, therefore increasing of fatigue properties is possible up to a certain level of hardening higher of which fatigue properties is being decreased.

3.1.1.4 Influence of depth of hardened zone

Relative gain of fatigue strength is being decreased with increasing of depth of hardened zone $\boldsymbol{l}$ under constant level of surface hardening K (**Figures 18** and **19**, **Table 3**). The higher level of K, the rather fatigue strength of alloy is being decreased with the rising of depth of hardened zone (**Figure 19**).

Solid solution hardening of surface layer of titanium alloys under conditions of thermodiffusion saturation by interstitial impurity (oxygen) can lead to the embitterment of hardened in this way layer and its brittle failure under conditions of the repeated loading. Because of this the investigations of fracture of samples were carried out after fatigue tests by rotating bending. Results of fractographical investigations of near-surface parts of fractures of titanium alloy VT1-0 samples with different levels of surface hardening after fatigue tests by rotating bending are presented in **Figure 20**.

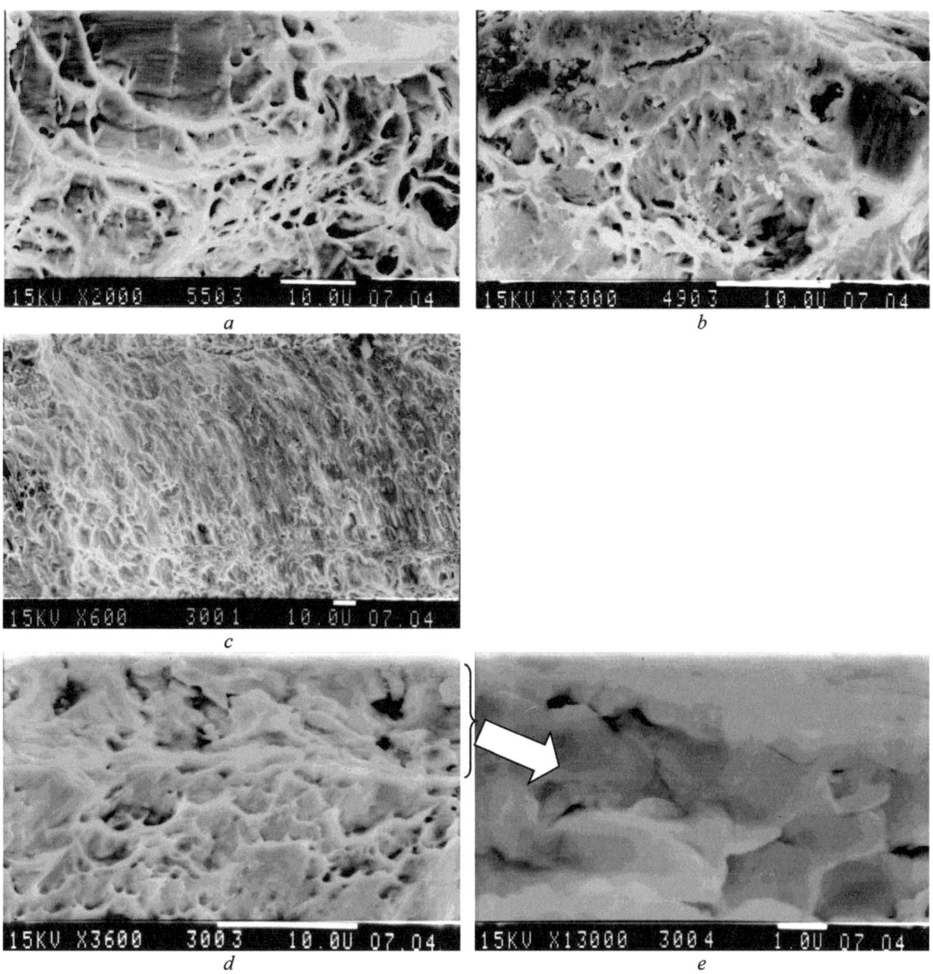

Figure 20.
Fractograms of near-surface part of fractures of titanium alloy VT1-0 samples with different levels of surface hardening after fatigue tests by rotating bending: (a) K = 70%, l = 30 μm, σ_{-1} = 310 MPa; *(b)* K = 50%, l = 70 μm, σ_{-1} = 285 MPa; *(c), (d), (e)* K = 70%, l = 70 μm, σ_{-1} = 260 MPa.

Cup-shaped fracture mode is fixed on the fractograms of near-surface layer of samples that testify in ductile failure (**Figure 20a–c**). That is, the total embrittlement of surface layer does not take place even if the level of fatigue strength is maximal (**Figure 7a**). Brittle mode of fracture of thin (~1...2 μm) near-surface layer is observed only on the specimen with K = 70%, l = 70 μm (**Figure 20d** and **e**). The time till origin of fatigue crack under alternate stress is being decreased providing that such layer exists on the samples and in turn the fatigue strength is being decreased. The decreasing of fatigue strength of alloy VT1-0 hardened by CTT with the increasing of depth of hardened zone l at a high level of surface hardening K is connected with this fact.

Thus, it can be concluded that for each of surface hardening level K of titanium alloy VT1-0 under conditions of thermodiffusion saturation in controlled gas medium, an optimal depth of hardened (gas saturated) zone l ensuring the highest level of fatigue characteristics exists. And vice versa for each depth of hardened zone, the optimal level of surface hardening exists. The aim of the next work stage is to search the optimal ratio of parameters K and l.

3.1.2 α-Alloy VT5

The parameters of regimes of thermodiffusion saturation of titanium alloy VT5 in the controlled gas medium containing oxygen and parameters of surface-hardened layer and cross section hardness distribution are presented in **Table 24**.

3.1.2.1 Influence of CTT on the fatigue properties of α-alloy VT5

The results of fatigue tests of samples of alloy VT5 after regulated surface hardening by thermodiffusion saturation in the gas medium are presented in **Figures 21–23** and **Table 25**.

It should be noticed that in the presented case, the character of dependence of fatigue strength σ_{-1} on the level of surface hardening K has a maximum level which depends on the depth of hardened zone l (**Figure 22**). The relative gain of fatigue strength $\Delta\sigma_{-1}$ of alloy VT5 reaches 20% under conditions K = 19%, l = 45–50 μm.

The fatigue strength is being decreased with the increasing of depth of gas-saturated layer under constant K (**Figure 23**).

It should be a supposition that under the analyzing of dependences presented in **Figures 22** and 23, the maximal gain of fatigue strength $\Delta\sigma_{-1}$ for alloy VT5 can be reached by the creation of gas-saturated layer of parameters $K \approx$ 45–55%, $l \approx$ 40–50 μm. Such parameters of gas-saturated layer can be determined as optimal parameters of the hardening of alloy VT5. It is the determination of optimal parameters of hardening to provide the highest gain of fatigue strength that is the aim of the next project step.

3.1.3 Near-α-alloy OT4-1 and (α + β)-alloy VT16

The regularities of the analogue presented above (see **Tables 26** and 27) are observed for the near α-alloy OT4-1 and (α + β)-alloy VT16. The sufficient effect of solid solution hardening is observed for alloy OT4-1: relative fatigue strength gain $\Delta\sigma_{-1}$ reaches 38% under relative gain of surface hardness K = 35%, l = 45–50 μm. It can be concluded that for titanium alloys with low or middle level of strength (VT1-0 and OT4-1 alloys), the positive effect of thermodiffusion saturation of metal surface layers by interstitial impurities is the highest: $\Delta\sigma_{-1}$ = 35–40%.

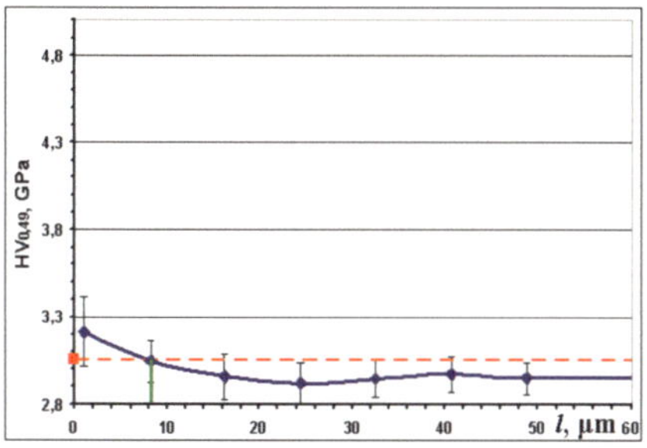

R0 – initial state: $K = 5\%$; $l = 10$ µm;
annealing $T = 800$ °C, $t = 2$ h, $P = 5\times10^{-5}$ Pa,
$I_{in} = 0.1\cdot10^{-3}$ Pa·dm^3·s^{-1} + polishing + annealing
$T = 600$°C, $t = 1$ h, $P = 5\times10^{-5}$ Pa,
$I_{in} = 0.1\cdot10^{-3}$ Pa·dm^3·s^{-1}

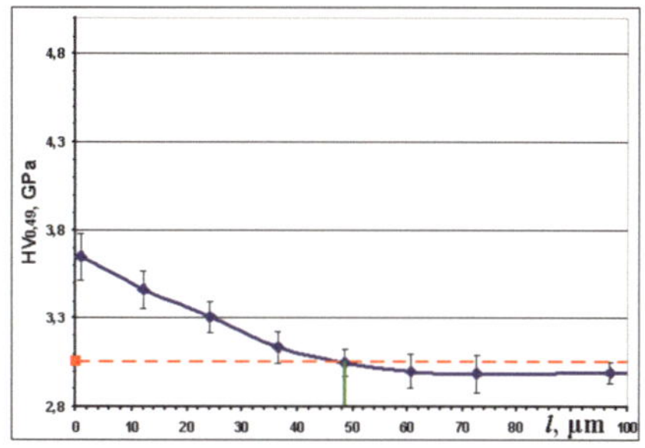

R1 - $K = 19\%$; $l = 50$ µm
R0 + annealing $T = 750$°C, $t = 5$ h, $P = 4\times10^{-2}$ Pa,
$I_{in} = 0.12\cdot10^{-3}$ Pa·dm^3·s^{-1}

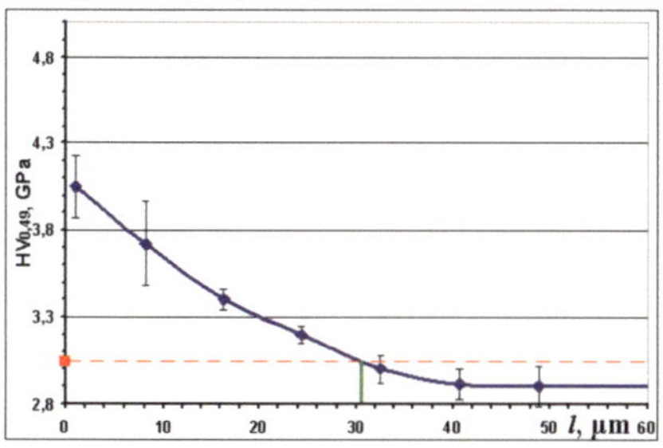

R2 - $K = 32$ %; $l = 30$ µm
R0 + annealing $T = 700$°C, $t = 5$ h,
$P = 1.3\times10^{-3}$ Pa, $I_{in} = 0.12\cdot10^{-3}$ Pa·dm^3·s^{-1}

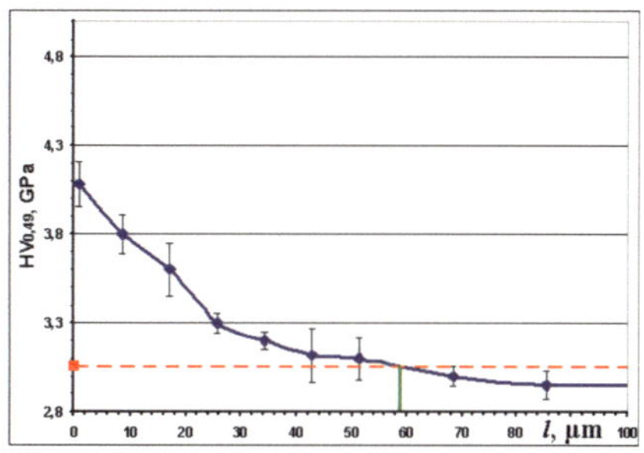

R3 - $K = 34$ %; $l = 60$ µm
R0 + annealing $T = 700$°C, $t = 5$ h, $P = 1.3\times10^{-3}$
Pa
+ annealing $T = 700$°C, $t = 5$ h, $P = 1.3\times10^{-2}$ Pa,
$I_{in} = 0.12\cdot10^{-3}$ Pa·dm^3·s^{-1}

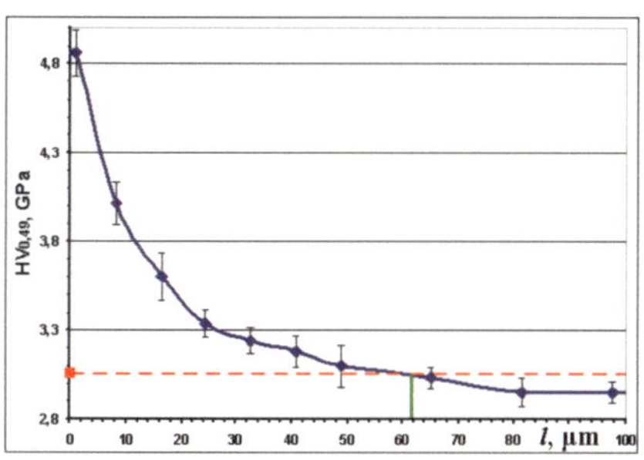

R4 - $K = 60\%$; $l = 60$ µm
R0 + annealing $T = 750$°C, $t = 6$ h,
$P = 6.6\times10^{-2}$ Pa, $I_{in} = 0.16\cdot10^{-3}$ Pa·dm^3·s^{-1}

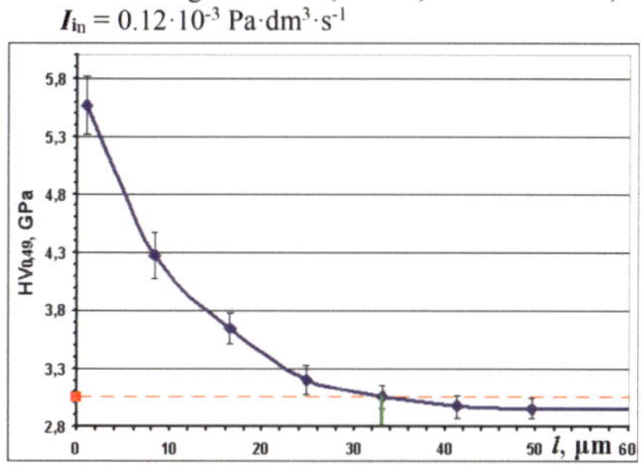

R5 - $K = 82\%$; $l = 30$ µm
R0 + annealing $T = 700$°C, $t = 5$ h,
$P = 6.6\times10^{-2}$ Pa, $I_{in} = 0.16\cdot10^{-3}$ Pa·dm^3·s^{-1}

Table 24.
Parameters of VT5 titanium alloy surface-hardened layers and thermodiffusion saturation's regimes.

The sensitivity of titanium alloys to the presence of gas-saturated layers is increased with the increasing of β-phase value. As a result the increasing of fatigue characteristics is attained under smaller values of relative gain ***K*** %. The value of fatigue strength gain is being decreased. Results of VT16 alloy tests can be a confirmation (**Table 27**).

Thus it can be concluded that for each level of surface hardening ***K*** of investigated titanium alloys VT1-0, VT5, OT4-1, and VT16 under conditions of thermodiffusion saturation in controlled gas medium, the optimal depth ***l*** of hardened (gas-saturated) zone exists which provides the highest level of fatigue

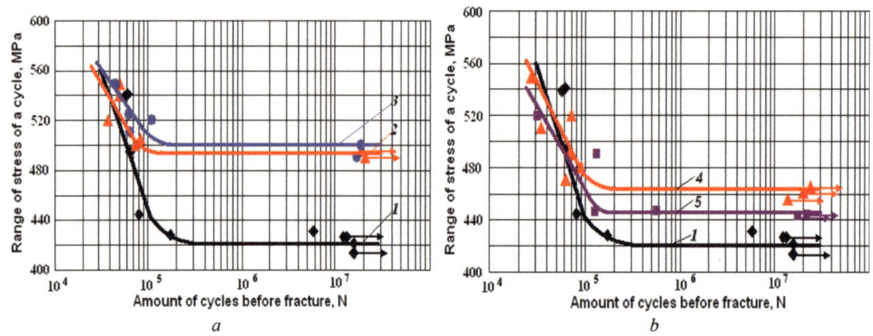

Figure 21.
*Fatigue curves of titanium alloy VT5, under rotating bending conditions, depending on level of surface hardening **K**, when depth of hardened (gas saturated) zone **l** is constant (a) **l** = 30–35 μm and (b) **l** = 60–65 μm: (1) initial state **K** = 5%; **l** = 5–10 μm; (2) **K** = 32%; (3) **K** = 82%; (4) **K** = 34%; (5) **K** = 60%.*

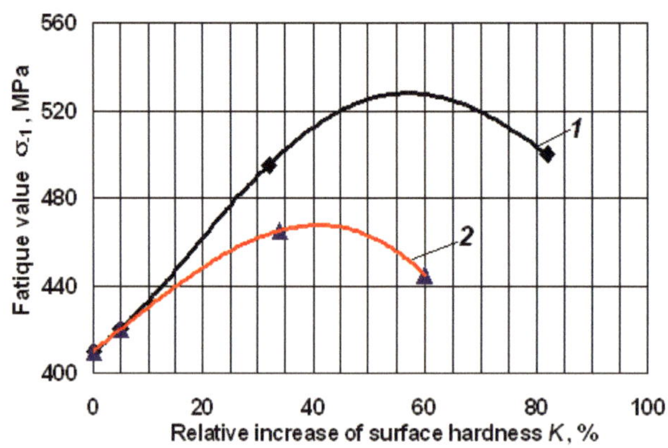

Figure 22.
*Fatigue strength of titanium alloy VT5, under rotating bending conditions, as a function of level of surface hardening when depth of hardened zone (gas saturated) is constant: (1) **l** = 30–35 μm (2) **l** = 60–65 μm.*

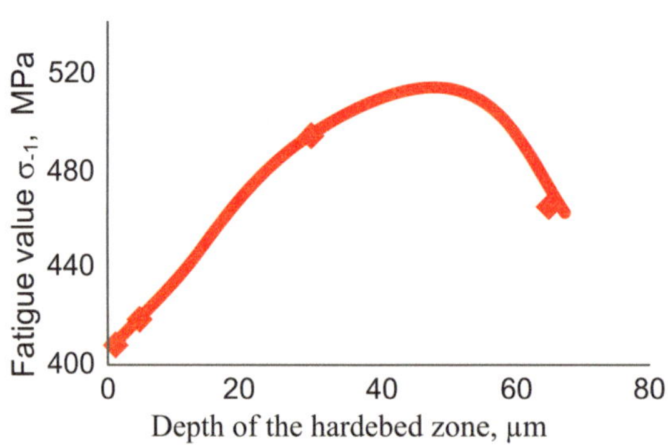

Figure 23.
*Fatigue strength of titanium alloy VT5, under rotating bending conditions, as a function of depth of hardened zone, when level of surface hardening is constant **K** = 30%.*

characteristics. In addition, vice versa for each depth of hardened zone, the optimal level of surface hardening exists. Furthermore, it can be forecasted that for each alloy the optimal relation between parameters ***K*** and ***l*** exists that provides absolutely maximal gain of fatigue strength. Such parameters of gas-saturated layers can be determined as optimal parameters of alloy hardening. The aim of the next step lies in the search of the optimal level of ***K*** and ***l*** parameters of the hardened zone.

#	Hardness of matrix H_O, MPa	Hardness of surface H_S, MPa	Relative gain of surface hardness K, %	Depth of hardened zone l, μm	Fatigue strength σ_{-1}, MPa	Relative gain of fatigue strength $\Delta\sigma_{-1}$, %
R0	3050 (initial state)	3200	5.0	5–10	≤420	0
R1	3050	3630	19.0	45–50	≤505	20.0
R2		4030	32.0	30–35	≤495	18.0
R3		4080	34.0	60–65	≤465	11.0
R4		4860	60.0	60–65	≤445	6.0
R5		5550	82.0	30–35	≤500	19.0

Table 25.
*Fatigue strength of titanium alloy VT5, under rotating bending conditions, depending on the level of surface hardening **K** and depth of hardened zone **l**.*

#	Hardness of matrix H_O, MPa	Hardness of surface H_S, MPa	Average relative gain of surface hardness K, %	Depth of hardened zone l, μm	Fatigue strength σ_{-1}, MPa	Relative gain of fatigue strength $\Delta\sigma_{-1}$, %
1.	2650 (initial state)	2850	4.0	5–10	≤335	0
2.	2650	3095	12.5	25–30	≤430	28.5
3.		3575	30.0	30–35	≤420	25.5

Table 26.
*Fatigue strength of titanium alloy OT4-1, under rotating bending conditions, depending on level of surface hardening **K** and depth of hardened zone **l**.*

#	Hardness of matrix H_O, MPa	Hardness of surface H_S, MPa	Average relative gain of surface hardness K, %	Depth of hardened zone l, μm	Fatigue strength σ_{-1}, MPa	Relative gain of fatigue strength $\Delta\sigma_{-1}$, %
1.	3000 (initial state)	3020	≤1	5–10	≤525	0
2.	3000	3145	5.0	100–120	≤580	10

Table 27.
*Fatigue strength of titanium alloy VT16, under rotating bending conditions, depending on level of surface hardening **K** and depth of hardened zone **l**.*

4. Conclusions

1. The kinetic parameters of interaction and regularities of solid solution hardening of titanium alloys VT1-0, VT5, OT4-1, and VT16 under conditions of thermodiffusion saturation in rarefied gas medium are determined.

2. It is shown that under the same conditions of saturation (T, τ, P), the hardened layers of various parameters (H, l) are formed on the titanium alloys. The monophase α-titanium alloys VT1-0 and VT5 and near-α-alloy OT4-1 are the most sensitive to the conditions of gas saturation: the gain of surface hardness and its gradient in the hardened layer increase sufficiently. With the increasing of β-phase (OT4 → VT16), changing of the parameters of CTT has less

influence on the hardness of surface layer, but the depth of the hardened zone is being increased with the increasing of the temperature and exposure time.

3. It is determined that solid solution hardening of titanium alloys VT1-0, VT5, OT4-1, and VT16 under conditions of thermodiffusion saturation by interstitial impurities in controlled gas medium leads to the increasing of fatigue strength under a definite ratio of hardened layers $\boldsymbol{K}$ and $\boldsymbol{l}$. The character of dependence of fatigue strength (σ_{-1}) of investigated alloys on the level of surface hardening ($\boldsymbol{K}$) has a maximum value which depends on the depth of hardened zone ($\boldsymbol{l}$). The positive influence of surface hardening on the fatigue characteristics is decreased under increasing of $\boldsymbol{l}$ when $\boldsymbol{K}$ is constant. The highest relative gain of fatigue strength ($\Delta\sigma_{-1}$) of samples with ChTT surface-hardened layers is marked for the low- and middle-strong alloys VT1-0 and OT4-1. Thus for alloy VT1-0, $\Delta\sigma_{-1} = 35\%$ under relative gain of surface hardness $\boldsymbol{K} = 70\%$ and $\boldsymbol{l} = 30$ μm. For the near-α alloy OT4-1, $\Delta\sigma_{-1} = 38\%$ under relative gain of surface hardness $\boldsymbol{K} = 35\%$ and $\boldsymbol{l} = 45$–50 μm.

4. The surface hardening of titanium alloy VT1-0 by thermodiffusion saturation (within the limits of parameters $\boldsymbol{K}$ and $\boldsymbol{l}$) provides saving of enough reserve of plasticity and does not influence the character of metal failure accordingly with the fractographical investigations. The failure of surface-hardened layer takes place by ductile plastic mechanism accordingly with fractographical investigations.

Author details

Vasyl Trush*, Viktor Fedirko and Alexander Luk'yanenko
Department of High Temperature Strength of Structural Materials in Gas and Liquid Metal Media, Karpenko Physico-Mechanical Institute of the National Academy of Sciences of Ukraine, Lviv, Ukraine

*Address all correspondence to: tvs1981@list.ru

References

[1] Maksimovich GG, Fedirko VN, Ya I, Pichugin SAT. Thermal Treatment of Titanium and Aluminum Alloys in Vacuum and Inert Media. Kiev: Science; 1987. (in Ukrainian)

[2] Weeks KE, Block FE. In: Arsentiev PP, editor. Thermodynamic Properties of 65 Elements, their Oxides, Halides, Carbides and Nitrides. M.: Metallurgy; 1965. (in Russian)

[3] Fromm E, Gebkhardt E. Gazy Gases and Carbon in Metals. Moscow, Russia: Metallurgiya; 1980. (in Russian)

[4] Zwicker U. Titanium and its Alloys. Moscow: Metallurgy; 1979. (in Russian)

[5] Maksimovich GG, Fedirko VN, Zima MN. Influence of residual gas pressure in a vacuum atmosphere on the oxidation and gas saturation of titanium alloys. 1988;**24**(1):101-104. Fiz.—chemical fur. materials

[6] Raychenko AI. Mathematical Theory of Diffusion in Applications. Kiev, Dumka: Sciences; 1981. (in Russian)

[7] GOST 25.502-79. Methods of Metals Mechanical Testing. Methods of Fatigue Testing. Moscow: Izdatelstvo Standartov, 1986. (in Russian)

[8] Fedirko VM, Pichugin AT, Luk'yanenko OH, Siryk ZO. Evaluation of the serviceability of products made of titanium alloys with gas-saturated layers. Materials Science. 1996;**32**(6): 688-693

[9] Glazunov SG, Moiseev VN. Titanium Alloys. Construction Titanium Alloys. Moscow: Metallurgiya; 1974. 368 p. (in Russian)

[10] Kolachev BA, Sadcov VV, Talalaev VD, et al. Vacuum Annealing Titanium Constructions. Moscov: Mashinostroenie; 1991. 224 p. (in Russian)

[11] Gorinin IV, Chechulin BB. Titanium in Machine-Building. Moscow: Mashinostroenie; 1990. 400 p. (in Russian)

[12] Pultsin NM, Afonin VK. Changes of characteristics of surface layer of titanium alloys under influence of vacuum medium and high temperatures. In: Some Questions of Titanium alloy's Properties Changes under Influence of Processing Conditions. Leningrad; 1971. pp. 17-24. (in Russian)

[13] PI 1.2.139-80. Vacuum Heat Treatment of Components and Assembly Units from the Deformable Titanium Alloys. NIAT–VIAM. 1980. 6p. (in Russian)

Section 4

Machining

Chapter 7

Sustainable Machining for Titanium Alloy Ti-6Al-4V

Imran Masood

Abstract

Sustainability achievement of difficult-to-machine materials is a major concern nowadays. Titanium alloy Ti-6Al-4V machined for dry, conventional and cryogenic cooling and surface finish is selected as response to assess machining sustainability through variables: cutting power, machining time, machining cost, material removal rate and cutting tool life. Results indicate that cryogenic cooling is more sustainable than dry and conventional cooling.

Keywords: cryogenic cooling, sustainable machining, Ti-6Al-4V

1. Introduction

In recent years, attention has made on achieving comprehensive strategy over sustainable manufacturing due to increased emission of CO_2 in environment and waste. This ultimately will improve industry's economy [1]. Machining technology is referring to implement the sustainability, which has potential to improve environmental performance and save money. The problem in implementing sustainability in production companies is due to short-term financial planning. However, long-term strategy is necessary for sustainable manufacturing.

The initiatives for the sustainable development are established at different levels, e.g., UN, OECD and National level, and are well positioned on macro level of production [2] but are lacking in implementation at shop-floor level.

Conventional machining processes using mineral-based cutting fluids tend to increase environmental problems. Attempts are in focus to attain sustainable products and processes that will eliminate the bad effects of mineral-based cutting fluids. Alternative technique of cooling by using liquid nitrogen is in focus to eliminate the adverse effects of mineral oils. The cost of machining is a major element of a mechanical industry. Cutting tools having long tool life are preferred over those with short tool life in order to reduce overall machining cost and increase productivity.

In aerospace industry, the components are usually made from superalloys. These alloys can withstand the high operating temperatures and extreme physical stresses. Out of these alloys, titanium is the most commonly used alloy. Ti-6Al-4V shows good results in application where high strength-to-weight ratio along with excellent corrosion resistance is required. This alloy is used in aircraft turbine engine components, structural parts of aircrafts, aerospace fasteners, high-performance automotive parts, marine applications, medical devices and sports equipment.

Milling is one of the mostly used machining operations in manufacturing industry. The situation becomes very difficult to handle when hard materials are required to go through milling process. The probability of tool wearing and damaging of surface roughness increases when high cutting forces and high temperatures are involved at interface of cutting tool and work material.

Much energy is utilized in a machining operation along with the wastage of chips in material removal process. Operational safety and environmental friendliness of any machining process play a vital role in sustainability of process. It is estimated that machining process contributes to the gross domestic product (GDP) of developed countries by about 5% of total the GDP [3].

The present trend is to create new and smaller products of higher quality in a shorter time and at a lower cost. There are many materials, which can be classified as 'difficult-to-machine'. Machining of such materials requires special cutting tools. Cutting parameters affect directly on the surface roughness, tool wear and machining time.

The world has now become energy conscious, and every industry is going to follow the health and safety requirements at all levels. The concept of sustainable manufacturing has now become the key focus, which deals with the economic, social, safety and environmental issues. Proper selection of cutting tool, coolant and process type is important for efficient and economic machining. Alternative solutions of dissipating the heat generated at chip-tool interface and cutting tool materials is in exploration since the last few years. Titanium alloys and other materials used in aerospace industry need special attention for their machining as these are difficult-to-machine materials.

Experimental data shows that cryogenic cooling is more sustainable than dry and conventional cooling process, and it gives the best results for tool life, surface finish of machined part, productivity with least impact on the environment, least energy cost and machining cost.

1.1 Problem related with conventional machining

Difficult-to-machine materials are heat resistant; therefore, in machining of such materials, huge amount of cutting tools and coolants is used. The cost of a machined part mainly involves the cost of tools used, electrical energy and coolant cost.

In conventional methodology, the parts are machined using mineral-based coolants. These coolants have disadvantages in worker's health and environment where it is disposed of. Titanium alloys and other nickel-based alloys are difficult-to-machine, and therefore a number of cutting tools are wasted in their machining increasing the overall cost of the product. Conventionally used coolants have many problems associated with the health of workers and environmental impact.

The problem becomes more critical when machining the difficult-to-machine materials where high temperatures are built at the tool-chip interface. Using of conventional mineral-based coolant to reduce the temperatures and frictions is creating other environmental problems because disposition of these lubricants is not only harmful for human life but also creating issues for aquatic organisms. Social impact of conventional lubricant and coolants is increasing health and safety problems of workers due to exposure to toxic chemicals. The working environments are being polluted increasing both the mist and noise levels. Industrial setups are required to adopt the sustainability principles in order to avoid increasing cost and environmental and safety issues. Alternative coolants are in exploration phase to replace with the conventional mineral-based coolants.

2. Sustainable manufacturing

Sustainability is defined as the ability to preserve, to keep or to maintain something. When something is sustainable it means that it is able to be kept and continued [4]. The three dimensions of sustainability (**Figure 1**) are environmental, social and economic, sometimes adding technology as the fourth one [5]. Initially sustainable development defined by Brundtland is reported in 1987 as 'the way for improving the well-being and quality of life for the present and future generation'. More precisely it was defined as 'to meet the needs of the present without compromising the ability of future generations to meet their own needs' [6].

Manufacturing contributes by about 22% of Europe's GDP, while 70% of jobs in Europe are directly related with manufacturing [7]. Energy consumption is a major factor in manufacturing industry which is based on electrical energy and oil. According to the action plan for energy efficiency, industrial production is responsible for about 18% of the total energy consumption of Europe [8].

Sustainable manufacturing is important for the manufacturing industry as it helps to cope with the increasing environmental regulations, meeting the customer requirements for better environmental performance, lowering the material and energy costs resulting to greener products. Sustainable manufacturing is the creation of products using such processes that minimize the environmental impacts, conserving natural resources and energy, are safe for the communities, employees and customers and are economically sound. It has become a business imperative as companies across the world are facing increased costs of energy and materials coupled with higher expectations of customers, communities and investors.

In sustainable manufacturing, the things are created in such a way (**Figure 2**) that is economically, environmentally, socially and safety wise feasible for both the manufacturer and user. Greater emphasis is over the environmental issue followed

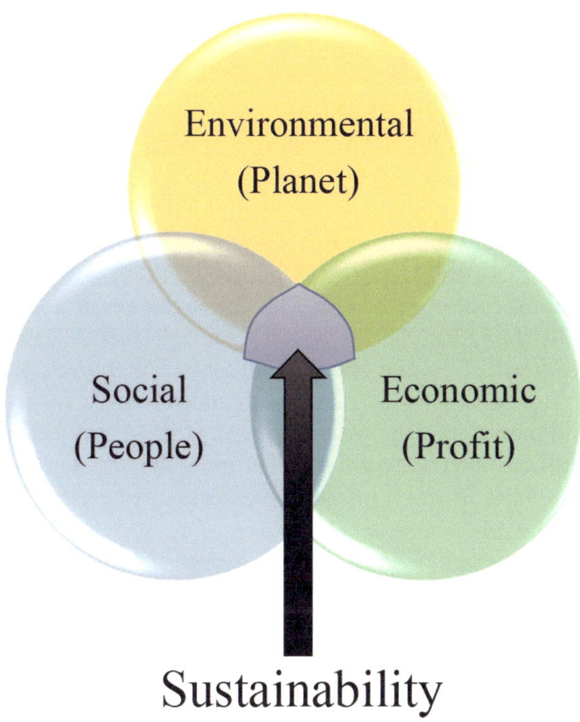

Figure 1.
Sustainability basic dimensions.

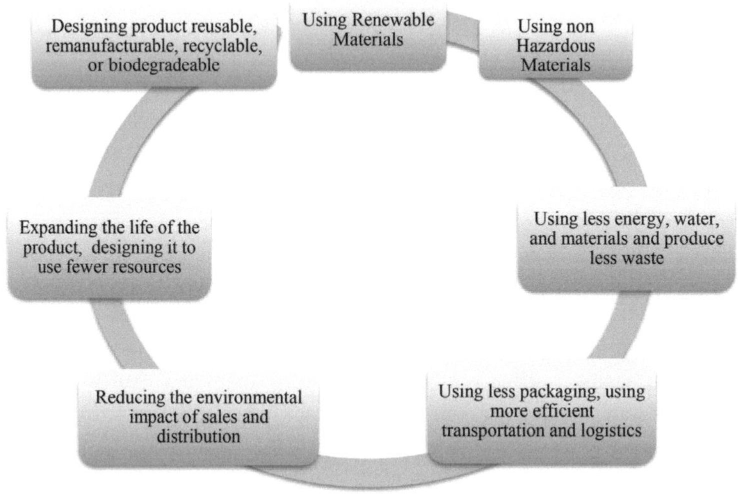

Figure 2.
Sustainable manufacturing aspects.

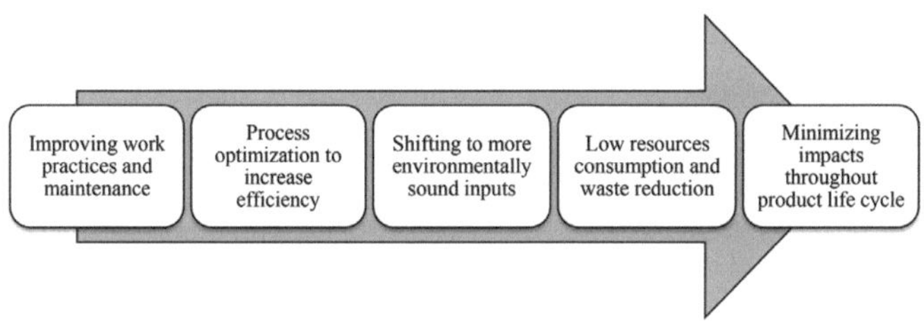

Figure 3.
Implementing sustainable manufacturing.

by the safety of workers and then economically. It leads the industry proactively to cleaner products and processes.

A process is said to be sustainable only when it has the lowest impact to the environment, is beneficial for society and is sound economically [9]. Hallmarks for the sustainable manufacturing are environmental friendly, lower machining costs, minimum energy consumption, personnel health, waste reduction and operational safety [10]. Globally the industries are striving for sustainable manufacturing by adopting advanced lubrication and cooling techniques, using vegetable oils or other environment-friendly cutting oils, selecting advanced tooling and accomplishment of advanced hybrid manufacturing processes, etc. [11]. Manufacturing industries of the USA, EU and other international countries are facing challenges to reduce the emissions of carbon dioxide and improve the efficiency of energy by making revolutions in technologies and processes [12].

The environment, social development, education, natural resources, poverty and inequality are examined for sustainable developments [13]. Sustainability has no destination or limits, but it has continuous improvement making the constant advances in company's overall sustainable performance. However spectrum of efforts can be followed as given by the United Nations Environment Programme [14] and OECD [15]. Sustainability is implemented by improving the work practices, optimizing processes, reducing resource consumption and minimizing impacts throughout the product life cycle (**Figure 3**).

2.1 Sustainable machining

Industrial trends are shifting from conventional to sustainable manufacturing principles. Such revolutions are outcome of diseases found in workers at shop floor, requirement of cost reduction for manufacturing and Government policies for environmental protection [16]. There is a need to make the machining systems sustainable in which processes are non-polluted, conserving both energy and natural resources, and economically sound and safe for employees [17]. Growing global competition and energy costs demand for such machining processes which are cheaper, using better utilization of resources and efficient usage of energy [18].

In the process of transforming inputs to outputs, the consumption of resources must be reduced to achieve sustainability. Refining the processes and machine tools are major factors for reducing the resources and energy consumption. The production systems are designed to support the continuous waste reduction, elimination or recycling. This can be achieved by less generation of waste; increasing recycling or re-usage; efficient usage of water, materials and energy; avoiding metal working fluids; and improving the management of lubricating oils, swarf and hydraulic oils. Sustainability will be gained by using the alternatives of cooling and lubrication fluids (CLFs) and dry machining using the coated cutting tools [19].

In a sustainable machining process, the tool life, productivity and effective utilization of resources will be increased, while the machining cost, machine cutting power and adverse effect of cooling and lubrication fluids will be decreased. The model of sustainable machining is presented in **Figure 4**.

Alternative cutting oils such as vegetable oils are renewable, environment-friendly and non-toxic in nature [20].

2.2 Conventional machining

Present investigations show that the cutting fluids are creating severe problems to health and environment [21]. The conventional fluids are considered very dangerous to the health and are rated out of top five hazardous to health [22]. Bulk use of conventional coolant in machining industries is causing increase in environmental damage [23].

Machining process contributes to worldwide economy, and it tends to become unsustainable when using such cooling and lubrication fluids which are oil based. These are made from mineral oils extracted from crude oil which is highly non-sustainable. Extract of crude oil is used to formulate the mineral oil which is converted to CLF. Although the vegetable oils are naturally derived, these are not used as CLF due to higher costs and reduced performance [24].

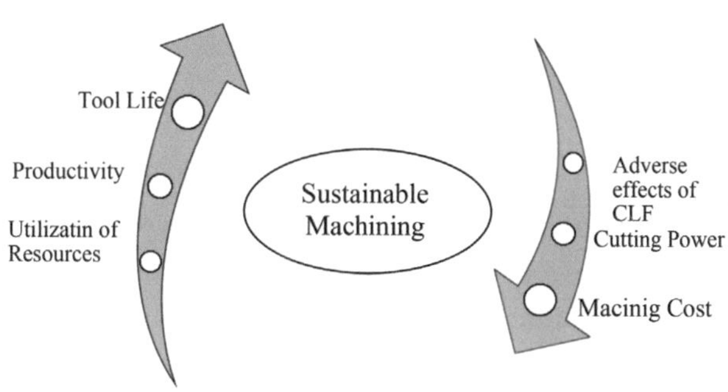

Figure 4.
Sustainable machining model.

CLFs are widely being used in metal cutting industry to counter the heat generated by machining besides that they have disadvantage of hazardous to health and environment. One of the most unsustainable elements in machining process is the use of cooling/lubrication fluid which is extracted from crude oils [25].

Cutting fluids are dangerous for health. In a report it is stated that about 80% of the skin diseases are due to the use of cutting fluids [26]. The machinists are facing the problem of skin and respiratory diseases due to metal working fluids [27]. Among the machinist over 1 million workers all over the world are facing toxic effects of cutting fluids, and majority of the cases are related to chest bronchitis [28]. According to a report [29], about 640 million gallon of coolant and lubrication fluid is used annually throughout the world. European Union estimated for metal working fluid and found that 3,20,000 tons is annually used and 66% of which is disposed-off after usage [30]. Used coolant in conventional machining has its adverse effects on the environment [31].

2.3 Cryogenic machining

Cryogenic machining is much safer than the conventional lubrication and coolants. Nitrogen gas has no hazards on life as about 79% of this already exist in air. Liquid nitrogen at cooling temperature is effective for cooling the cutting edge during machining of hard materials as cutting temperature exceeds 200°C [32]. It is a new technique of cooling the cutting zone and part during the machining at high speed and temperatures with cryogenic CLF. The coolant is nitrogen which is liquefied at −196°C and is safe, non-corrosive and non-combustible gas. This gas evaporates leaving no contaminates with part, operator, machine tool, and chips; thus disposal cost is eliminated. Mostly cryogenic CLFs are applied in the machining of superalloys.

The cryogenic machining process is more beneficial and more sustainable in terms of safety, clean and environment-friendly machining. Due to minimization in changeover time, productivity also increases. Tool life is increased due to low abrasion rate and chemical wear. Improvement is observed in the surface quality without the degradation in its mechanical/chemical properties.

For implementation of cryogenic machining at industrial level, investigations are required about the tool wear and tool life using cryogenic cooling [33]. Application of cryogenic machining at shop-floor level will be transitioning towards the sustainable machining and will promote the development of optimization for cryogenic fluid delivery with mass flow and controlled pressure.

In cryogenic machining the cryogenic fluid is directly applied on the cutting tip of the tool. This flow is manageable to be controlled against flow and pressure which makes it more economic than conventional fluids. N_2 gas is used as a cooling medium in cryogenic machining and is harmless to the health. This process increases the tool life and helps in productivity improvement, surface integrity improvement, chip breakability enhancement, reduction in built-up edge and burr formation [34–37].

In comparison of cryogenic cooling with conventional cooling and lubrication process, it is clear that the cost of power required for pumping of cooling and lubrication fluid is eliminated. The cost of cleaning CLF from the machined part becomes zero. Alternates of cutting fluid like N_2, O_2 and CO_2 have been used and compared to wet and dry machining and found that fine surface finish obtained with increased flow rates and pressure of gases [38].

Compressed air as coolant was used for machining of optical glass and found that low cutting forces are observed as compared with diamond drilling [39].

Experiments performed using liquid nitrogen in turning process of titanium alloy with modified tools resulted in improved tool life, surface finish and reduced cutting temperature of 65% and reduced cutting forces [40]. Experiments were also performed to check the machinability by considering the surface roughness. Ti-6Al-4V was machined in dry, wet and cryogenic conditions to observe the surface roughness. Surface finish is found consistently better with cryogenic than with dry and wet [41].

Different gases have been used as a coolant like CO_2, air, argon and nitrogen [42]. Experimental results conducted on Steel AISI 1040 using CO_2, O_2 and N_2 show that best surface finish is achieved using CO_2 then oxygen and nitrogen. Cryogenically compressed air was used for investigating the chip temperature, cutting force and the chip formation during the turning of Ti-6Al-4V [43].

Sustainability in machining can be assured by reduction in energy consumption for machining processes, minimizing waste (less generation of waste and increasing the recycling of waste).

Benefits of using the sustainable machining cover increasing MRR without increasing wear rate of the cutting tool, decreasing the tool changeover time increasing productivity and improving the machined surface without degradation which results in the presence of chemical coolants.

2.4 Machining issues of Ti-6Al-4V alloy

The most commonly used materials in aerospace industry are nickel alloys, titanium alloys and Co-Cr alloys [44]. Thermal conductivity of such materials is low, and therefore temperature observed at the cutting zone is extremely high. These facts have called for sustainability in machining and finding the alternate to conventional oil-based CLF as cooling and lubrication [45–48].

Higher temperatures are observed in the high-speed machining (HSM) that result in high temperatures at cutting tool and part interface. In the reports it is given that tooling cost is about 4% of the total machining costs and coolant/lubrication cost is about 15% of total machining cost [49]; therefore huge sustainability gain is possible by avoiding CLF and using high-performance coated cutting tools [50]. Titanium alloy (Ti-6Al-4V) is referred to as difficult-to-machine material. It has low thermal conductivity due to which very high temperatures occur during the machining at the cutting point. Its mechanical properties are very good for load-carrying applications due to which it is mostly used in the commercial and military aircrafts. **Figure 5** shows the comparison of machining difficulty level with other common materials.

This alloy, also called Grade 5 titanium, shows good results when used in applications where good mechanical and thermal properties along with good strength-to-weight ratio are the primary objectives. Due to its good results in strong environments and resistance to corrosion, it is also used in petroleum industries, nuclear reactors, turbine blades, marine applications and medical implants. Demand of

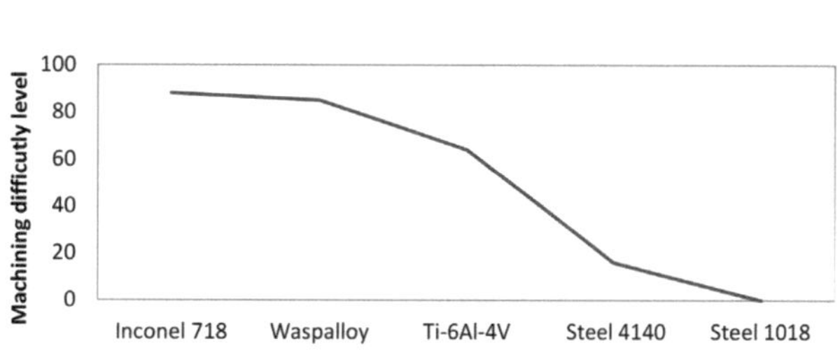

Figure 5.
Machining difficulty level of Ti-6Al-4V.

titanium parts is extensively increasing for industry of aircrafts such as Boeing 787. Preparation of titanium parts requires much cost for machining operations.

Due to low thermal conductivity, titanium has poor machinability. Titanium alloys can be used at temperature of 600°C. Titanium is related to a group of hard materials like nickel alloys, ceramics and cobalt-chromium alloys. It is important to cool down the cutting tool temperature in order to improve the cutting tool life, especially in the case when machining the materials with low thermal conductivity like Ti-6Al-4V alloy [51].

Coolant in conventional machining is harmful to aquatic organism and may cause long-term adverse effects in the aquatic environment. It is harmful to the respiratory system and can cause slight irritation making the environment contaminant. Repeated exposure may cause skin dryness. Nitrogen gas is harmless to environment and worker's health. Therefore, cryogenic machining nullifies the exposure to toxic chemicals making it safer for both workers and environment.

2.5 Machining problems of difficult-to-machine materials

In machining of difficult-to-machine materials like Ti-6Al-4V, excessive tool wear and heat are produced making the surface quality poor [52]. Alternative solutions of dissipating the heat generated at chip-tool interface and cutting tool materials are in exploration since the last few years. The main reasons for rapid tool wear are building of high cutting temperatures. In machining of hard and difficult-to-machine materials, the conventional CLF (oil-based) does not effectively decrease the cutting temperatures, and therefore tool life is not increased. It is due to the fact that the coolants do not access the chip-tool interface which is under high cutting temperature and vaporize close to the cutting edge. Due to this phenomenon, the conventional CLF becomes ineffective for machining the materials with low thermal conductivity and high shear strength.

Dry machining is not recommended at high-speed machining of difficult-to-machine materials. Such materials are used in aerospace industry and are capable of bearing high operating temperature like in jet engines.

2.6 Surface finish for metals

The quality of machined parts can be ensured by measuring the surface roughness. The quality of a surface with low roughness value is good over the surface having greater value of roughness. Surface finish is the important characteristic of precise devices as poor surface finish results in the problems of malfunctioning, geometric inaccuracy and excessive wear [53]. Surface finish and dimensional accuracy of a part greatly affect during its useful service life. Obtaining better surface finish of microstructures is in focus nowadays [54]. Failure of components commonly occurs as a result of poor surface of parts; therefore getting good surface finish is too much important. Researchers paid much attention towards getting good surface-finished parts using optimization techniques.

Surface roughness is mostly influenced by the feed rate, tool geometry, tool wear, chatter, tool deflection, cutting fluids and properties of working material. Other different kinds of factors (**Figure 6**) can affect the surface roughness.

Most of the researchers have used the machining parameters in their work to find the response over the surface roughness. For example, the large nose radius of cutting tool will produce better surface finish than the small nose radius [55]. The feed rate plays also its role in surface finish that the smaller feed rate yields better surface finish.

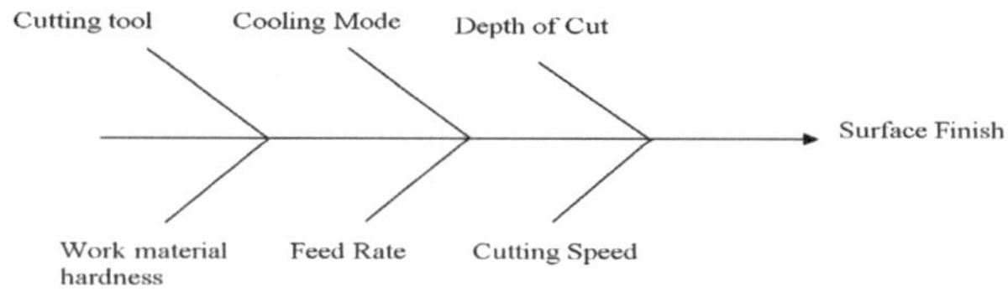

Figure 6.
Parameters that affect surface roughness.

2.7 Machining cost

Cost of a machined part mainly involves the cost of cutting tools, electrical energy and labour and coolant cost. High machining cost of titanium alloy Ti-6Al-4V has made it important to ensure longer tool life by selecting the favourable cutting conditions [56]. Present competitive trends of manufacturing are focused on generating the products with low cost and high quality. The cost of machining was computed based on the machining time, while total cost was calculated adding the machine cost, setup cost, material cost and non-productive costs [57].

A cost estimation model has been proposed in [58] for optimization of machining cost which includes material cost, tool cost and overhead and labour cost. In this proposed model, if the desired cost-effective results are not achieved, then the feedback is given to a designer for modifications. The feasible process parameters including cutting speed, feed rate and depth of cut are selected to attain optimum results. Constraints of cutting tool specification, tolerances, cost, time, machining sequence and required surface finish are taken into consideration.

3. Sustainability assessment for machining of Ti-6Al-4V

Sustainability of a machining process refers to the impact on environment, power consumption and safe for operator, which were satisfied in the experimental works as the cost of tool was reduced in the cryogenic cooling and it also impact on time saving for tool changing and setup time, which result in increasing productivity. An advantage of cutting in cryogenic process is evaporating back of cooling gas (Nitrogen) into air, which ensures the healthy environment for workers.

In experimental work [59], face milling of hardened Ti-6Al-4V at 55 HRC was carried for dry, conventional and cryogenic cooling conditions. Experiment model was designed using software design expert and technique of central composite design (CCD) was selected. Surface finish as response was checked and compared for each scenario of cooling. Resulting values of surface finish were compared based on iso-response technique, and the cutting power, cutting time, material removal rate, machining cost and cutting tool life were calculated as given in following sustainability parameters:

Cutting power: it was found that the cutting power required in cryogenic machining is 61.9% less than cutting power required in dry machining.

Machining cost: it is found that electricity cost is 47.55% less than dry machining and 14.22% less than conventional machining.

Adverse effects of CLF: the adverse effects of conventional coolants are reduced by replacing the coolant with N_2 gas. Corresponding cutting power and machining cost are also reduced.

Machining time: machining time for cryogenic machining is 15.12% less and for conventional machining it is 12.51% less than dry machining.

Material removal rate: the material removal rate is 81.12% more for cryogenic than dry machining.

Tool life: cutting tool's life is 5.2 times more for cryogenic cooling which indicates that the waste in the form of damaged tools is reduced.

4. Conclusions

From the experimental results, it is concluded that cryogenic machining is recommended for Ti-6Al-4V. Results are satisfying sustainability for eco-friendly impact on the environment, reducing tooling and energy cost. Efforts can be made to switch from conventional machining to cryogenic machining which would be beneficial in reducing machining costs, health risks along with fine surface surface.

The minimum value of surface finish can be obtained by the cryogenic machining using the coated carbide cutting tools. The cutting tool will not be damaged by cryogenic cooling ensuring both the sustainability and cost saving. Comparison of cutting power, cutting time, electricity cost, coolant cost, machine operating cost and material removal rate (**Figure 7**) for nearly identical response of surface finish shows that the cryogenic machining is more sustainable than others.

The results of tool life describe that cutting tool will survive for longer time in cryogenic cutting conditions than dry and conventional, resulting low cost of tool for the machining processes. Similarly, the cost evaluation resulted in low machining cost for cryogenic cooling as compared to dry and conventional. Cryogenic machining is more affordable and economic as there is no cost of pumping coolant, very low cost for cutting tool inserts and labour.

Cutting tool inserts were found damaged in dry machining, whereas very low wear was found in conventional, and no wear was found in cryogenic machining.

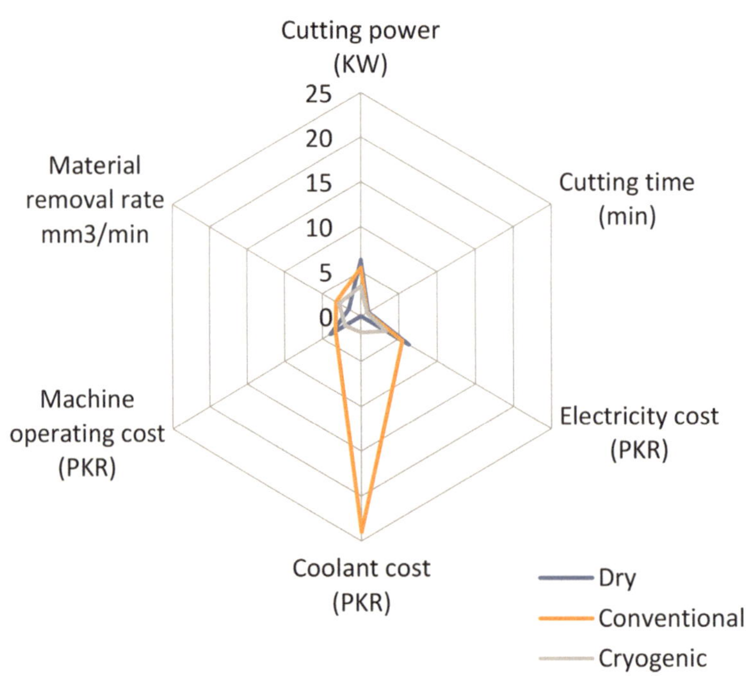

Figure 7.
Average response values calculated for nearly identical surface finish.

The coolant used in conventional machining has its adverse effects on worker's health and environment, while the nitrogen gas is harmless. The tool wear rates are also high for dry and conventional. Summarizing all findings, it can be stated that cryogenic machining supports the sustainable machining.

Author details

Imran Masood
Industrial Engineering Department, University of Engineering and Technology, Taxila, Pakistan

*Address all correspondence to: imranmasood76@gmail.com

References

[1] Yuan C, Zhai Q, Dornfeld D. A three dimensional system approach for environmentally sustainable manufacturing. CIRP Annals-Manufacturing Technology. 2012;**61**(1): 39-42

[2] Jovane F, Yoshikawa H, Alting L, Boër C, Westkamper E, Williams D, et al. The incoming global technological and industrial revolution towards competitive sustainable manufacturing. CIRP Annals-Manufacturing Technology. 2008;**57**(2):641-659

[3] Lee G, Badrul O. Optimization for sustainable manufacturing based on axiomatic design principles: A case study of machining processes. Advances in Production Engineering & Management. 2014;**9**(1):31-43

[4] Ciceri ND, Garetti M, Sperandio S. From product end-of-life sustainable considerations to design management. In: Advances in Production Management Systems. New Challenges, New Approaches. Springer; 2009. pp. 152-159

[5] Baud R. The concept of sustainable development: Aspects and their consequences from a social-philosophical perspective. In: Presented at the YES Youth Encounter on Sustainability Summer Course Material; Braunwald, Switzerland. 2008

[6] B Commission. Our Common Future: Report of the World Commission on Environment and Development. Oxford (UK); 1987

[7] EU-Commission. Manufacture—A Vision for 2020 Assuring the Future of Manufacturing in Europe. Luxembourg: Office for Official Publications of the European Communities; 2004. ISBN 92-894-8322-9

[8] Communication from the Commission. Action Plan for Energy Efficiency: Realising the Potential. Brussels; 19.10.2006 COM (2006) 545 final

[9] Gupta K, Laubscher R, Davim JP, Jain N. Recent developments in sustainable manufacturing of gears: A review. Journal of Cleaner Production. 2016;**112**(1):3320-3330

[10] Davim JP. Sustainable Manufacturing. London; Hoboken, NJ: ISTE; John Wiley; 2010

[11] Fratila D. Environmentally friendly manufacturing processes in context of transition to sustainable production. Comprehensive Materials Processing. 2014;**8**(1):163-175

[12] Park C-W, Kwon K-S, Kim W-B, Min B-K, Park S-J, Sung I-H, et al. Energy consumption reduction technology in manufacturing—A selective review of policies, standards, and research. International Journal of Precision Engineering and Manufacturing. 2009;**10**(5):151-173

[13] Abdalla H, Patel S. The performance and oxidation stability of sustainable metalworking fluid derived from vegetable extracts. Proceedings of the Institution of Mechanical Engineers, Part B: Journal of Engineering Manufacture. 2006;**220**(12):2027-2040

[14] Ashford N. Government Strategies and Policies for Cleaner Production. France: United Nations Environment Programme, Industry and Environment; 1994

[15] OECD, editor. Eco-Innovation in Industry: Enabling Green Growth. Paris, France: OECD; 2009

[16] Jayal A, Badurdeen F, Dillon O, Jawahir I. Sustainable manufacturing: Modeling and optimization challenges at the product, process and system levels.

CIRP Journal of Manufacturing Science and Technology. 2010;**2**(3):144-152

[17] Dornfeld D. Sustainability of machining. In: CIRP Encyclopedia of Production Engineering. Berlin, Heidelberg: Springer; 2014. pp. 1204-1207

[18] Deiab I. On energy efficient and sustainable machining through hybrid processes. Materials and Manufacturing Processes. 2014;**29**(11-12):1338-1345

[19] Westkämper E. Manufacturing Systems and Technologies for the new Frontier. London: Springer; 2008. pp. 11-14

[20] Debnath S, Reddy MM, Yi QS. Environmental friendly cutting fluids and cooling techniques in machining: A review. Journal of Cleaner Production. 2014;**83**(1):33-47

[21] Yildiz Y, Nalbant M. A review of cryogenic cooling in machining processes. International Journal of Machine Tools and Manufacture. 2008; **48**(9):947-964

[22] Jawahir I, Dillon O. Sustainable manufacturing processes: New challenges for developing predictive models and optimization techniques. In: Proceedings of the First International Conference on Sustainable Manufacturing; Canada; 2007. pp. 1-19

[23] Zhang S, Li J, Wang Y. Tool life and cutting forces in end milling Inconel 718 under dry and minimum quantity cooling lubrication cutting conditions. Journal of Cleaner Production. 2012;**32**: 81-87

[24] Herrmann C, Hesselbach J, Bock R, Zein A, Öhlschläger G, Dettmer T. Ecologically benign lubricants—Evaluation from a life cycle perspective. CLEAN-Soil, Air, Water. KGaA, Weinheim: WILEY-VCH Verlag GmbH & Co.; 2007;**35**(5):427-432

[25] Pusavec F, Krajnik P, Kopac J. Transitioning to sustainable production—Part I: Application on machining technologies. Journal of Cleaner Production. 2010;**18**(2):174-184

[26] Lawal S, Choudhury I, Nukman Y. Application of vegetable oil-based metalworking fluids in machining ferrous metals—A review. International Journal of Machine Tools and Manufacture. 2012;**52**(1):1-12

[27] Hannu T, Suuronen K, Aalto-Korte K, Alanko K, Luukkonen R, Järvelä M, et al. Occupational respiratory and skin diseases among Finnish machinists: Findings of a large clinical study. International Archives of Occupational and Environmental Health. 2013;**86**(2): 189-197

[28] Marksberry P. Micro-flood (MF) technology for sustainable manufacturing operations that are coolant less and occupationally friendly. Journal of Cleaner Production. 2007; **15**(10):958-971

[29] Brockhoff T, Walter A. Fluid minimization in cutting and grinding. Abrasives, Journal of Abrasives Engineering Society. 1998;**10**(11):38-42

[30] Abdalla H, Baines W, McIntyre G, Slade C. Development of novel sustainable neat-oil metal working fluids for stainless steel and titanium alloy machining. Part 1. Formulation development. The International Journal of Advanced Manufacturing Technology. 2007;**34**(1-2):21-33

[31] Heisel U, Lutz M, Spath D, Wassmer RA, Walter U. Application of minimum quantity cooling lubrication technology in cutting processes. Production Engineering. 1994;**II**(1): 49-54

[32] Pusavec F, Kopac J. Achieving and implementation of sustainability

principles in machining processes. Journal of Advances in Production Engineering and Management. 2009;**3**: 58-69

[33] Shokrani A, Dhokia V, Newman ST. Investigation of the effects of cryogenic machining on surface integrity in CNC end milling of Ti-6Al-4V titanium alloy. Journal of Manufacturing Processes. 2016;**21**:172-179

[34] Ghosh R, Zurecki Z, Frey JH. Cryogenic machining with brittle tools and effects on tool life. In: ASME 2003 International Mechanical Engineering Congress and Exposition. Washington, DC, USA; 2003. pp. 201-209

[35] Dhar NR, Kamruzzaman M, Khan M, Chattopadhyay A. Effects of cryogenic cooling by liquid nitrogen jets on tool wear, surface finish and dimensional deviation in turning different steels. International Journal of Machining and Machinability of Materials. 2006;**1**(1):115-131

[36] Da Silva FJ, Franco SD, Machado ÁR, Ezugwu EO, Souza AM Jr. Performance of cryogenically treated HSS tools. Wear. 2006;**261**(5):674-685

[37] Olteanu EL, Bîşu CF, Tănase I. Determination of power consumption in milling. Scientific Bulletin, Series D, University Politehnica of Bucharest. 2013;**75**(4):211-220

[38] Çakır O, Kıyak M, Altan E. Comparison of gases applications to wet and dry cuttings in turning. Journal of Materials Processing Technology. 2004; **153**:35-41

[39] Zhang C, Cong W, Feng P, Pei Z. Rotary ultrasonic machining of optical K9 glass using compressed air as coolant: A feasibility study. Proceedings of the Institution of Mechanical Engineers, Part B: Journal of Engineering Manufacture. 2014;**228**(4):504-514

[40] Dhananchezian M, Kumar MP. Cryogenic turning of the Ti-6Al-4V alloy with modified cutting tool inserts. Cryogenics. 2011;**51**(1):34-40

[41] Rotella G, Dillon O Jr, Umbrello D, Settineri L, Jawahir I. The effects of cooling conditions on surface integrity in machining of Ti6Al4V alloy. The International Journal of Advanced Manufacturing Technology. 2014;**71** (1-4):47-55

[42] Kim S, Lee D, Kang M, Kim J. Evaluation of machinability by cutting environments in high-speed milling of difficult-to-cut materials. Journal of Materials Processing Technology. 2001; **111**(1):256-260

[43] Sun S, Brandt M, Dargusch M. Machining Ti-6Al-4V alloy with cryogenic compressed air cooling. International Journal of Machine Tools and Manufacture. 2010;**50**(11):933-942

[44] Ezugwu E. Key improvements in the machining of difficult-to-cut aerospace superalloys. International Journal of Machine Tools and Manufacture. 2005;**45**(12):1353-1367

[45] Dudzinski D, Devillez A, Moufki A, Larrouquere D, Zerrouki V, Vigneau J. A review of developments towards dry and high speed machining of Inconel 718 alloy. International Journal of Machine Tools and Manufacture. 2004; **44**(4):439-456

[46] Ezugwu E, Bonney J. Effect of high-pressure coolant supply when machining nickel-base, Inconel 718, alloy with coated carbide tools. Journal of Materials Processing Technology. 2004;**153-154**(1):1045-1050

[47] Kramar D, Kopac J. High performance manufacturing aspect of hard-to-machine materials. Journal of Advances in Production Engineering & Management (APEM). 2009;**4**(1-2):3-14

[48] Su Y, He N, Li L, Iqbal A, Xiao M, Xu S, et al. Refrigerated cooling air cutting of difficult-to-cut materials. International Journal of Machine Tools and Manufacture. 2007;**47**(6):927-933

[49] Weinert K, Inasaki I, Sutherland J, Wakabayashi T. Dry machining and minimum quantity lubrication. CIRP Annals—Manufacturing Technology. 2004;**53**(2):511-537

[50] Skerlos SJ, Hayes KF, Clarens AF, Zhao F. Current advances in sustainable metalworking fluids research. International Journal of Sustainable Manufacturing. 2008;**1**(1-2):180-202

[51] Park K-H, Yang G-D, Suhaimi M, Lee DY, Kim T-G, Kim D-W, et al. The effect of cryogenic cooling and minimum quantity lubrication on end milling of titanium alloy Ti-6Al-4V. Journal of Mechanical Science and Technology. 2015;**29**(12):5121-5126

[52] Shokrani A, Dhokia V, Newman ST. Environmentally conscious machining of difficult-to-machine materials with regard to cutting fluids. International Journal of Machine Tools and Manufacture. 2012;**57**(1):83-101

[53] Kim W-B, Lee S-H, Min B-K. Surface finishing and evaluation of three-dimensional silicon microchannel using magnetorheological fluid. Journal of Manufacturing Science and Engineering. 2004;**126**(4):772-778

[54] Kim W-B, Park S-J, Min B-K, Lee S-J. Surface finishing technique for small parts using dielectrophoretic effects of abrasive particles. Journal of Materials Processing Technology. 2004;**147**(3): 377-384

[55] Groover MP. Fundamentals of Modern Manufacturing: Materials Processes, and Systems. USA: John Wiley & Sons; 2007

[56] Jaffery S, Mativenga P. Assessment of the machinability of Ti-6Al-4V alloy using the wear map approach. The International Journal of Advanced Manufacturing Technology. 2009;**40** (7-8):687-696

[57] Shehab E, Abdalla H. Manufacturing cost modelling for concurrent product development. Robotics and Computer-Integrated Manufacturing. 2001;**17**(4): 341-353

[58] Gayretli A, Abdalla H. A prototype constraint-based system for the automation and optimization of machining processes. Proceedings of the Institution of Mechanical Engineers, Part B: Journal of Engineering Manufacture. 1999;**213**(7):655-676

[59] Masood I, Jahanzaib M, Haider A. Tool wear and cost evaluation of face milling grade 5 titanium alloy for sustainable machining. Advances in Production Engineering & Management. 2016;**11**(3):239-250

Chapter 8

The Comparison of Cutting Tools for High Speed Machining of Ti-6Al-4V ELI Alloy (Grade 23)

Chakradhar Bandapalli, Bharatkumar Mohanbhai Sutaria and Dhananjay Vishnu Prasad Bhatt

Abstract

Green technology is one of the major aspects in order to reduce the global pollution content from manufacturing industries. There is a need to investigate the different available tools for high-speed micromilling process of advanced alloys to achieve desired surface finish without traditional coolants. In this chapter, tool wear investigation of uncoated and PVD-coated AlTiN, TiAlN tungsten carbide end mills in high-speed micro-end milling of alpha + beta Ti-6Al-4V ELI titanium alloy (Grade 23) under dry cutting conditions was presented. A comparison for machining performance with the three tools is reported. Cutting force analysis was done under the considered machining input parameters for evaluating the tool condition. Tool wear observation was done by SEM analysis. EDX analysis was performed to know the material constituents and wear mechanisms on the cutting tool tip. It is found that diffusion, oxidation, adhesive and abrasive wear mechanisms were the major phenomena taking place on the cutting edge of micro end mills. From the comparison of cutting tools for machining Grade 23 titanium alloy, it was found that TiAlN tools performed better than AlTiN and uncoated tungsten carbide tools.

Keywords: green technology, micro-end milling, PVD-coated AlTiN and TiAlN tungsten carbide end mills, tool wear, titanium alloys, and uncoated tungsten carbide end mills

1. Introduction

Titanium and its alloys materials are broadly used in applications like air frame components, medical implants/devices, surgical instruments, ballistic armour, space vehicles/structures, missile components, navy ship components, chemical processing equipment, hydrocarbon refining/processing, hydrometallurgical extraction/electrowinning offshore hydrocarbon production, desalination, brine concentration/evaporation, power generation, automotive, mining, railways and sporting goods. The large variety of application is due to its desirable properties, mainly the relative high strength combined with low density and enhanced corrosion resistance. In terms of biomedical applications, the properties of interest are biocompatibility, corrosion behaviour, mechanical behaviour, process ability and availability [1–5].

IntechOpen

In machining of brittle and ductile materials, selection of available tools with different grades is a complex matter. Economics and quality of the machining are dependent on tool wear. Evaluation and measurement of tool wear in micromilling are challenging compared to the conventional machining process. In high-speed micromilling based on the surface finish requirement on the desired work material, tools with smaller diameter, that is, two flute and four flute end mills, are used for superfinishing operation. Two, four and more flute tools having larger tool diameter are used for the roughing operation. Tool wear occurrence in four flutes is lower than two flutes as the cutting and load bearing capacity is higher for four flutes in roughing and super finishing operations. In super finishing, at high spindle speeds for a slotting operation which is single pass cutting, the usage of micro-end mills is limited, maybe two to three slots. Burr formation, cutting forces, surface roughness and acoustic emission signals observation and analysis information provide the tool wear prediction in high-speed micro-end milling. Tool material properties and machining parameters decide the tool wear formation. Tool wear of a different kind takes place in micromilling because of small cutting edge, microstructure variation, difference in work and tool material phases, deformed chips, wrong rake angle tools selection shape and a number of flutes, friction and stress induced [6–9] on the tool.

Komanduri and Reed [10] investigated the cutting performance of carbide grades and new cutting geometry in turning operation of titanium alloys. They observed that prolonged tool life in machining Ti alloys can be obtained at high clearance angle and high negative rake angle. Kitagawa et al. [11] investigated the temperature and wear of cutting tools in high speed machining of Ti-6Al-6V-2Sn and found that temperature plays the major role for tool wear during machining. According to Jawaid et al. [12], CVD-coated tools performed well during face milling of Ti-6Al-4V than PVD tools. They observed nonuniform flank wear pattern on both the tools and found that coating delamination, diffusion, attrition, adhesion wear mechanisms were responsible factors. Liu et al. [13] investigated cutting forces and surface quality in micromilling of TC4 titanium alloy. They found that surface quality of machined surface is prone to the influence of burrs and residual chips. Nouari et al. [14, 15] investigated CVD tools and uncoated tools performance in machining titanium alloy Ti-6242S and they observed almost equal performance of both tools. They observed similar physical phenomenon while machining as mentioned by Jawaid et al. [12]. Rahman et al. [16] presented a review on high-speed machining of titanium alloys especially in turning and milling operations. They discussed the performance of coated and uncoated tungsten carbide tools, polycrystalline diamond (PCD) tools, cubic boron nitride (CBN) tools, and binderless cubic boron nitride (BCBN) tools in terms of cutting forces generation and tool wear. They found that BCBN tools performed well in high-speed machining conditions and traditional tools in moderate cutting speed conditions. Ginta et al. [17] investigated tool wear morphology and chip segmentation in end milling of titanium alloy Ti-6Al-4V using uncoated WC-Co inserts. They also performed modelling and optimization of tool life and surface roughness. They observed abrasion/attrition, plastic deformation and diffusion wear processes. They found that combination of high cutting speed and feed substantially increases the stress near the nose and flank zone, generates high temperature and encourages high wear rate.

Schueler et al. [18] investigated the burr formation mechanisms and surface characteristics in micro-end milling Ti-6Al-4V and Ti-6Al-7Nb titanium alloys. Large areas were machined to observe the microstructure on the surface and the influence on surface quality. Up milling and down milling at the sidewalls were compared. They found that down milling is better than up milling. Arrazola et al. [19] investigated and compared the cutting forces, tool wear and chip geometry in the machining of ($\alpha + \beta$) Ti-6Al-4V and near-beta (β) Ti555.3 titanium alloys. They found that adhesive and diffusion wear on cutting tools when machined with both the grades of titanium alloys. Specific feed force and cutting force are higher for Ti555.3 alloy than Ti-6Al-4V alloy. Chip formation observed was segmented with and without adiabatic shear zones in Ti-6Al-4V alloy and narrow adiabatic shear bands for Ti555.3 alloy. Malekian et al. [20] investigated tool wear monitoring in micromilling processes to avoid the failure of tools during the considered machining conditions. Tool edge radius and wear were observed and measured using vision system and as well as gathered sensor signals of acoustic emission, acceleration and force data. These data were interpreted offline using adaptive neuro-fuzzy inference ystem and compared with experimental wear results that were agreeable. Smith et al. [21] investigated surface quality and tool wear in micromilling of Ti-6Al-4V using monocrystalline CVD diamond cutting edges with preferential crystallographic orientation. The analyses of tool wear and workpiece surface quality proved that monocrystalline CVD diamond cutting edges with preferential crystallographic orientation along rake and clearance faces can be successfully utilised for interrupted cutting operations (i.e., micromilling) of alloys which react with diamond, such as those based on titanium. Zhang et al. [22] investigated the cutting forces and tool wear variations during high-speed micro-end milling of titanium alloy ($\alpha + \beta$) Ti-6Al-4V using uncoated cemented tungsten carbide tools. They found that due to adhesion, abrasion and diffusion process, tool wear takes place and cutting force component Fy, in the considered experiment, has a positive relationship with the tool wear propagation.

Ozel et al. [23–25] carried out experimental investigation and finite element simulation with CBN tools and uncoated tools in the micromilling of Ti-6Al-4V. They found that larger the feed rate, the higher the burr formation, surface roughness, temperature generation, cutting forces and tool wear. CBN tools had less tool wear and temperature formation than uncoated tools. Wyen et al. [26] investigated the influence of the cutting-edge radius on surface integrity in slot milling of Ti-6Al-4V with different edge radius tools. As the cutting temperature and kinematics influence the up and down machining processes, they thoroughly researched on cutting edge radius owing to the temperature generation. They found that down milling is better than up milling for surface roughness and burr formation which gradually increase with increasing cutting-edge radius from the measurements of residual stress and compressive stress generation. Durul and Ozel [27] presented review on machining-induced surface integrity in titanium and nickel alloys. They reported detailed performance of the different tools, tool wear behavior, burr formation, surface topography and FEM simulations. Bajpai et al. [28] investigated surface quality and burr formation in HSMEM of Ti-6Al-4V and it was found that as cutting speed, feed rate and depth of cut increased, then smoother surface finish can be achieved. Burr formation is increased due to increment in depth of cut. Hou et al. [29] investigated the influence of cutting speed on tool wear, flank temperature and cutting force in macro-end milling of titanium alloy Ti-6Al-4V using PVD-coated TiN/TiAlN and uncoated tungsten carbide tools. They found that high cutting forces were generated when cutting speed is increased and increment in mean flank temperature for the coated cutting tools, whereas it is almost constant for uncoated tools. At higher cutting speeds, no abrasion and fatigue wear were observed for uncoated tools and in contrast with coated tools. Kim et al. [30] discussed the machining input parameters influence and found that spindle speed and feed rate were the most influencing factors for generating cutting forces and burr formation on Ti-6Al-4V. Pervaiz et al. [31] presented a review on influence of tool materials on machinability of titanium- and nickel-based alloys. Ceramic, PCD, CBN, boronized carbide tools and high pressurised coolant supply show good results for machining. They observed that in experimental studies that high pressurised coolant supply reduces cutting temperature and improves chip breakability which results in improved tool life. Increase in the pressure of coolant supply also improves tool life. Notch wear was reduced by increasing coolant supply pressure when machining titanium alloys.

Hassanpour et al. [32] investigated the cutting force, microhardness, surface roughness and burr size in micromilling of Ti-6Al-4V using minimum quantity lubrication. They found that cutting speed and feed per tooth significantly affect the surface roughness. Bandapalli et al. [33] have investigated the influence of machining parameters in high-speed micro-end milling of commercially pure titanium Grade 2 and found that cutting forces and surface roughness formation increases at high spindle speed by increasing feed rate and depth of cut. Ghani et al. [34] investigated the wear mechanism of uncoated carbide cutting tool in milling of aluminium metal matrix composite (AlSi/AlN MMC), PVD-coated TiAlN/AlCrN tool in milling of Inconel 718 and uncoated tool in turning of Ti-6Al-4V ELI. They found that tools failed primarily on two main areas of the flank and rake faces for cutting the Inconel 718 and titanium alloy. Wear such as crater, nose wear, abrasion, notching, fracturing and cracking were observed. In machining AlSi/AlN MMC, the tools mainly failed due to the uniform flank wear that was caused by abrasion.

However, research findings related to considered machining parameters in high-speed micro-end milling (HSMEM) of titanium alloys for comparison and evaluation of uncoated, PVD-coated TiAlN and AlTiN tools are inadequate. The purpose of this research work is to evaluate the performance of tools like uncoated and physical vapour deposition (PVD)-coated TiAlN and AlTiN tools in terms of tool wear formation in HSMEM of alpha + beta Ti-6Al-4V ELI titanium alloy (Grade 23). The goal was to improve the quality or productivity of the specific machining process based on empirical experiments using a variety of speeds, feeds and depth of cut. This work is categorised into four sections, first with an introduction. Experimental details were discussed in Section 2. Results and discussion were presented in Sections 3 and 4 and finally Section 5 is presented with conclusions.

2. Experimental details

Standardised experimental tests are carried out on high-speed micromachine setup as shown in **Figure 1**. The work material considered was Ti-6Al-4V ELI titanium alloy (Grade 23) of dimensions 60 × 40 × 4 mm as shown in **Figure 2**. Chemical composition and properties of the work material are shown in **Tables 1** and 2. Experimental tests conducted are shown in **Table 3**. Two flute uncoated tungsten

Figure 1.
High speed micromachine.

Figure 2.
Ti-6Al-4V ELI titanium alloy (Grade 23).

Properties	Metric
Tensile strength	860 MPa
Yield strength	790 MPa
Poisson's ratio	0.342
Elastic modulus	113.8 GPa
Shear modulus	44.0 GPa
Elongation at break	15%
Hardness Rockwell	35
Density	4.43 g/cm^3
Melting point	1604–1660°C

Table 1.
Mechanical and physical properties of Ti-6Al-4V ELI titanium alloy (Grade 23).

Element	wt%
Titanium, Ti	88.09–91
Aluminium, Al	5.5–6.5
Vanadium, V	3.5–4.5
Iron, Fe	≤0.25
Carbon, C	≤0.080
Nitrogen, N	≤0.030
Hydrogen, H	≤0.0125
Other	≤0.50
Total	100

Table 2.
Chemical composition of Ti-6Al-4V ELI titanium alloy (Grade 23).

carbide end mills and physical vapour deposition-coated TiAlN and AlTiN tungsten carbide end mills of diameter 500 μm supplied by IND-SPHINX Axis tools were used in this work as shown in **Figure 3**. The total length of each machined slot was 12 mm. Tool over hang length considered was 18 mm. The width of cut is 500 μm as the process is slot milling operation. Cutting edge radius of tool identified is 2.71 μm through SEM. Coating of the material as specified by the IND-SPHINX Axis tools is

Exp. no	Spindle speed (rpm)	Feed (μm/ tooth)	Depth of cut (mm)	Cutting force (N)		
				Uncoated	TiAlN	AlTiN
1	30,000	2	0.02	0.37	0.34	0.33
2	30,000	5	0.02	0.39	0.34	0.31
3	30,000	8	0.02	0.41	0.36	0.34
4	30,000	2	0.06	0.46	0.63	0.62
5	30,000	5	0.06	0.57	0.68	0.67
6	30,000	8	0.06	0.69	0.81	0.66
7	30,000	2	0.1	0.75	0.76	1.5
8	30,000	5	0.1	0.82	0.88	1.55
9	30,000	8	0.1	1.01	1.16	1.56
10	70,000	2	0.02	0.09	0.14	0.13
11	70,000	5	0.02	0.21	0.21	0.21
12	70,000	8	0.02	0.37	0.29	0.29
13	70,000	2	0.06	0.15	0.26	0.22
14	70,000	5	0.06	0.31	0.31	0.38
15	70,000	8	0.06	0.47	0.40	0.41
16	70,000	2	0.1	0.21	0.35	0.57
17	70,000	5	0.1	0.43	0.46	0.54
18	70,000	8	0.1	0.56	0.58	0.63
19	110,000	2	0.02	0.12	0.13	0.22
20	110,000	5	0.02	0.31	0.28	0.28
21	110,000	8	0.02	0.49	0.34	0.28
22	110,000	2	0.06	0.16	0.15	0.19
23	110,000	5	0.06	0.37	0.27	0.36
24	110,000	8	0.06	0.56	0.37	0.37
25	110,000	2	0.1	0.18	0.16	0.25
26	110,000	5	0.1	0.46	0.34	0.61
27	110,000	8	0.1	0.67	0.41	0.77

Table 3.
Resultant cutting forces for uncoated, coated TiAlN and AlTiN WC tools.

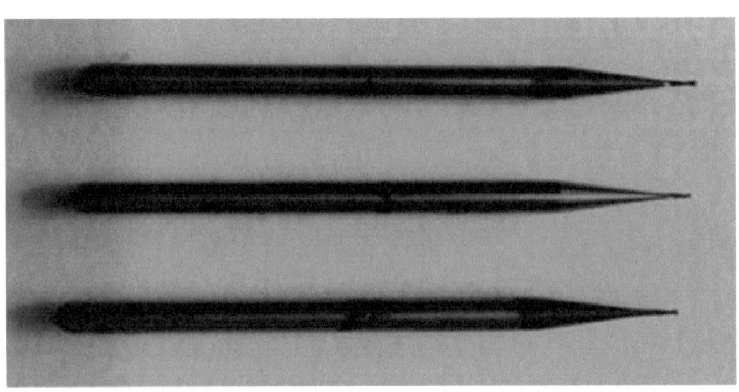

Figure 3.
Micro-end mills.

2–6 μm. Rake angle of the tool is +5°. Static run-out of the tool was measured as 3 μm. No structural vibrations were observed under considered machining conditions. Chip thickness observed is about 3 μm. Machining time was 10–12 s for 30,000 rpm, 6–8 s for 70,000 rpm and 3–5 s for 110,000 rpm.

Parametric experiments, full factorial design 3^3 = 3 factors, each with three levels, 3^3 = 27 total runs were conducted for determining the effect of the process parameters on the cutting force and tool wear. Three levels of tool rotation speed—30,000, 70,000 and 110,000 rpm, that is, cutting velocity of 47, 110 and 173 m/min, three levels of feed rate—2, 5 and 8 μm/tooth and three levels of depth of cut—0.02, 0.06 and 0.1 mm were selected in these experiments. Two sets of experimentation, that is, one original and one repetition were performed in which total number of machined slots are—162 (27 × 2 × 3 tool types). Six uncoated tools for 54 experiments, 6 TiAlN tools for 54 experiments and 6 AlTiN tools for 54 experiments were used in this experimentation. End mill was changed with new one after machining nine slots on the workpiece for verification and observation of cutting forces and tool wear. Coated tools were selected because they provide high wear resistance, withstand mechanical and thermal shock, plastic deformation, act as barrier towards wear formation, reduce subsurface defects on workpiece by generating less heat, high hot hardness, chemical inertness to reduce development of built-up edge and occurrence of coating delamination, less burr formation and ability to withstand high cutting forces. In accordance with the above view, uncoated and PVD-coated TiAlN and AlTiN tungsten carbide end mills were considered in the present research work. Tool wear was examined using HITACHI-S3400N scanning electron microscope (SEM) equipped with energy dispersive spectroscope (EDS).

3. Results and discussion

3.1 Cutting force analysis

Micro-end milling is one of the most commonly used machining processes and has more complex geometry due to its rotating tool, multiple cutting edges and intermittent cutting action. Cutting forces are the main cause of the deformations of machine tool structures and workpieces resulting in form errors and tolerance violations. Although they may affect the structural components of a machine tool distributed in a large space, cutting forces are generated in a very small area at the work-tool interface. A cutting origin was set through a CCD camera because the tool diameter was extremely small and direct origin setting with naked eye was difficult and could result in significant errors. Cutting forces in three directions F_x, F_y and F_z measured using tool dynamometer, and from the signal analyser, the forces were interpreted in the computer. Resultant cutting forces were calculated by Eq. (1). Experimental results of resultant cutting forces are shown in **Table 3**.

$$F_R = \sqrt{F_x^2 + F_y^2 + F_z^2} \qquad (1)$$

3.1.1 Micromilling with uncoated tools

At 30,000 rpm, if depth of cut and feed is varied, then resultant cutting force increased by 63% as shown in **Figure 4(a)**. At 70,000 rpm, if depth of cut and feed is varied, then resultant cutting force increased by 87% as shown in **Figure 4(b)**. At 110,000 rpm, if depth of cut and feed is varied, then resultant cutting force increases by 83% as shown in **Figure 4(c)**.

3.1.2 Micromilling with TiAlN tools

At 30,000 rpm, if depth of cut and feed is varied, then resultant cutting force increases by 81.7% as shown in **Figure 4(a)**. At 70,000 rpm, if depth of cut and feed is varied, then resultant cutting force increases by 85.14% as shown in **Figure 4(b)**. At 110,000 rpm, if depth of cut and feed is varied, then resultant cutting force increases by 78% as shown in **Figure 4(c)**.

3.1.3 Micromilling with AlTiN tools

At 30,000 rpm, if depth of cut and feed is varied, then resultant cutting force increases by 82.6% as shown in **Figure 4(a)**. At 70,000 rpm, if depth of cut and feed is varied, then resultant cutting force increases by 87% as shown in **Figure 4(b)**. At 110,000 rpm, if depth of cut and feed is varied, then resultant cutting force increases by 85% as shown in **Figure 4(c)**.

From the experimentation, it is observed that the resultant cutting force increase might be due to the increment in feed rate and depth of cut, leading to more chip formation, high contact between tool and chip, high-temperature formation by shearing and ploughing action reducing the yield strength of work material. It was observed that when spindle speed is increased, while the depth of cut and feed rate is kept at low, then cutting force generated was less. When depth of cut and feed rates are reduced, the chip load encountered in the process becomes the same order of magnitude as the grain size of many alloys. The cutting-edge radius of the end mill is comparable in size to the chip thickness. As a result, no chip is formed when the chip thickness is below the minimum chip thickness and instead part of the work material plastically deforms under the edge of the tool and the rest elastically recovers. This change in the chip formation process, known as minimum chip thickness effect and the associated material elastic recovery cases, increased cutting forces and surface roughness. Observed irregularity on the machined surface due to the plastic side flow and burr formation seem to suggest that the tool workpiece interaction is more likely to be elastic-plastic in nature. Due to the increment in feed rate, depth of cut and cutting speed, the tool wear and cutting-edge distortion took place that leads to more cutting force requirement to remove the material. At 30,000 rpm, uncoated tools performed better than both the coated tools. Uncoated tools produced less cutting force compared to coated tools because of tools stability and resistance to wear. At 70,000 rpm, 0.02 and 0.06 mm depth of cut with feed 2–8 μm/tooth, the uncoated tool required more cutting force to remove the material rather than both the coated tools. At 70,000 rpm, 0.1 mm depth of cut with feed 2–8 μm/tooth, the uncoated tool is found to generate less cutting force than both the coated tools. At 110,000 rpm, coated TiAlN tools are better performed than coated AlTiN and uncoated tools, which is due to increase in cutting temperature in the shear zone, thus reducing the yield strength of the workpiece material, chip thickness and tool-chip contact length. If coated TiAlN and coated AlTiN tools are compared, then performance of TiAlN tools is better. Depending on the particular parameters during machining processes, uneven phenomenon (sudden tool or spindle vibration, sudden workpiece movement) occurs, leading to increase of cutting forces. SEM and EDS analyses are performed to identify the wear and its formation mechanism on two flutes of each tool, that is, uncoated and PVD-coated AlTiN, TiAlN tungsten carbide end mills as shown in **Figures 5** and **6**. Tool wear on particular end mill is observed using SEM after machining nine slots for each set of spindle speed as shown in **Figures 7–12**. Tool wears measured on each cutting flute in two series of experiments have been considered and average values are reported as shown in **Figure 13**. Coated TiAlN tool wear was high at spindle speed

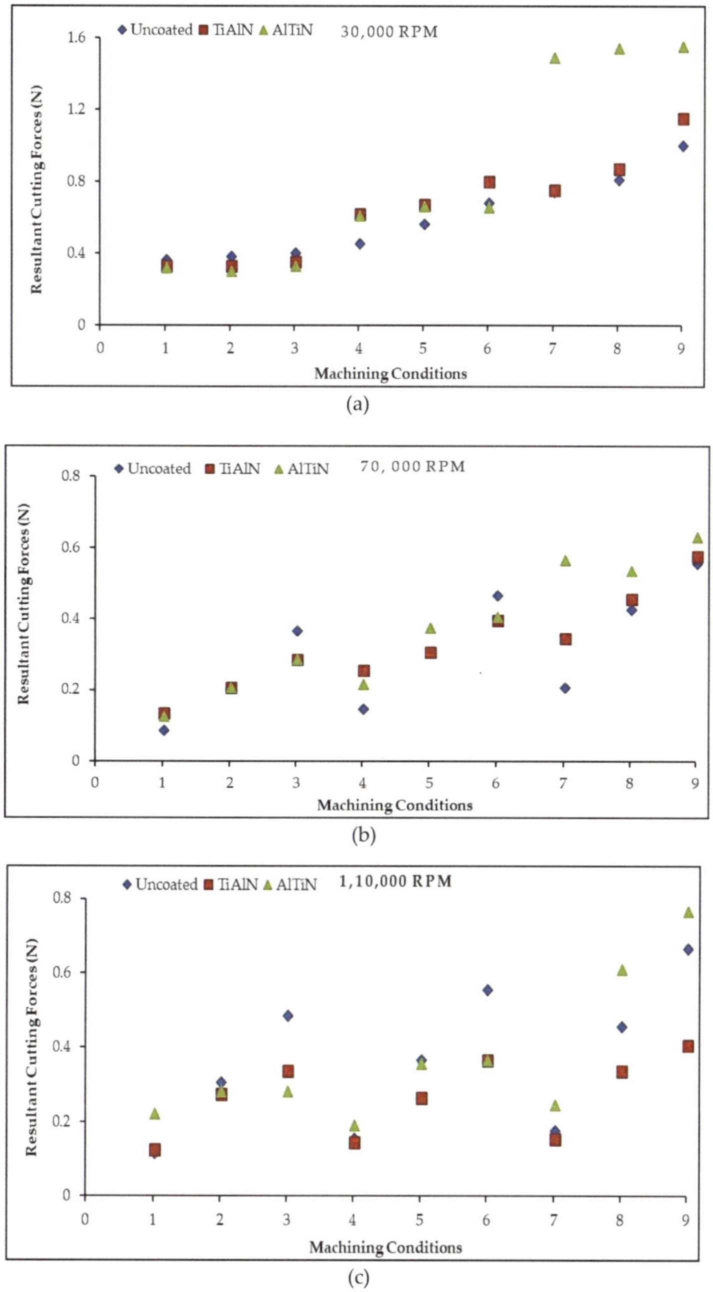

Figure 4.
Comparison of (a) uncoated, (b) coated TiAlN and (c) AlTiN WC tools for resultant cutting force (N) at independent spindle speed (rpm), feed (μm/tooth), depth of cut (mm).

30,000 rpm machining conditions than uncoated and coated AlTiN tools. Uncoated tools produced less wear than both the coated tools at spindle speed 70,000 rpm machining conditions. Coated TiAlN tools produced less wear than AlTiN-coated and uncoated tools at spindle speed 110,000 rpm machining conditions. Tool wear is increased for all the tools when the feed rate and depth of cut for the considered spindle speed is increased. Adhesive wear, edge chipping, flaking, coating delamination and measured tool wear that were observed under SEM at different machining conditions are shown in **Figures 7–12**.

Figure 5.
SEM image of tool edge for uncoated, coated TiAlN and AlTiN WC tools with marked microzone for EDS analysis.

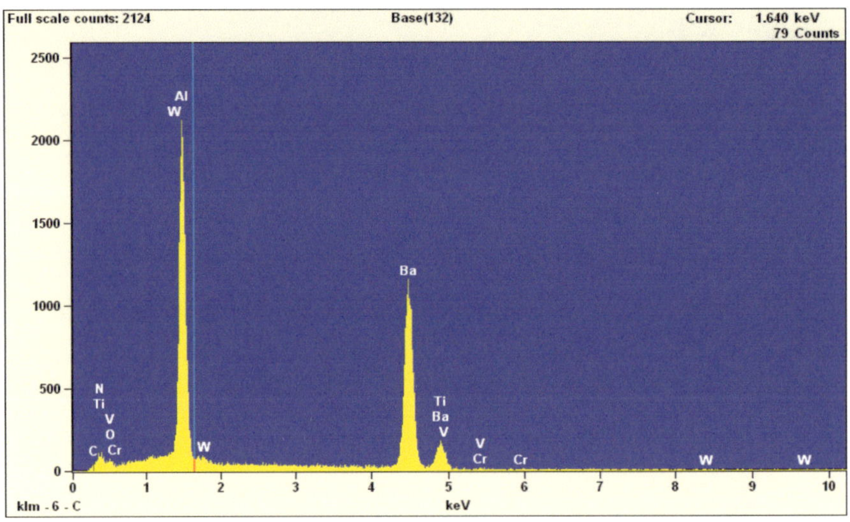

Figure 6.
EDS spectrogram.

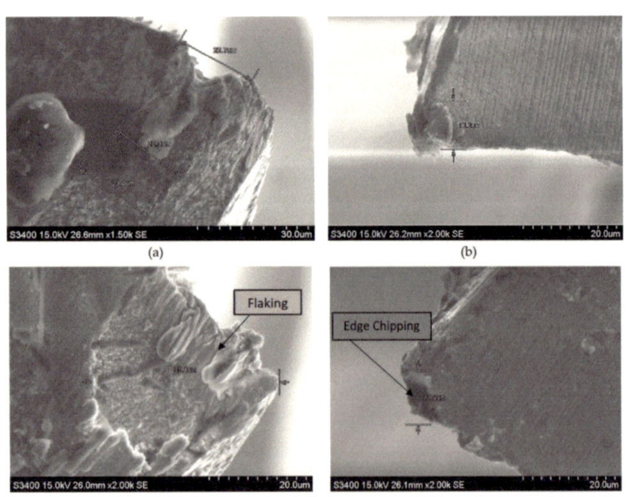

Figure 7.
(a–d) Tool wear observation for uncoated tools.

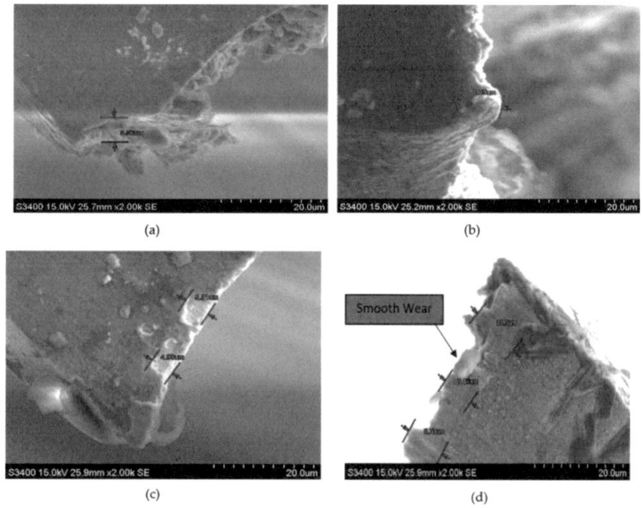

Figure 8.
Coated TiAlN tool wear at (a and b) 30,000 rpm and (c and d) 70,000 rpm.

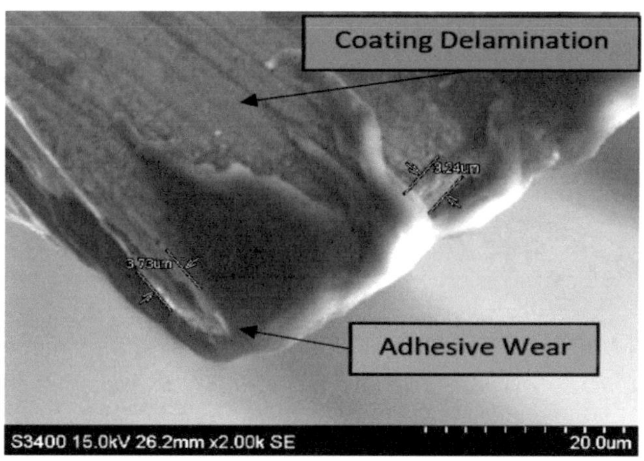

Figure 9.
Coated TiAlN tool wear at 110,000 rpm.

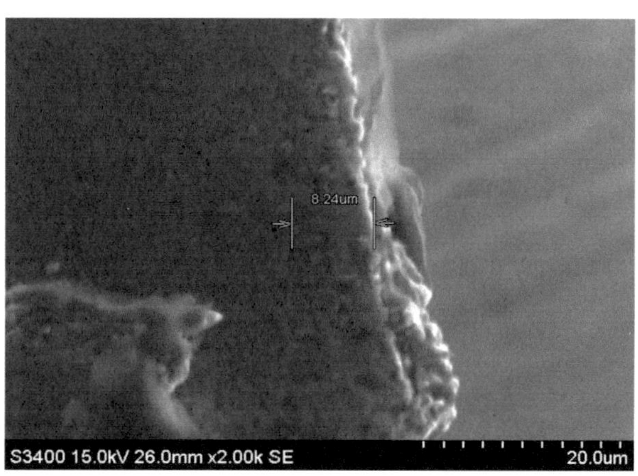

Figure 10.
Coated AlTiN tool wear at 30,000 rpm.

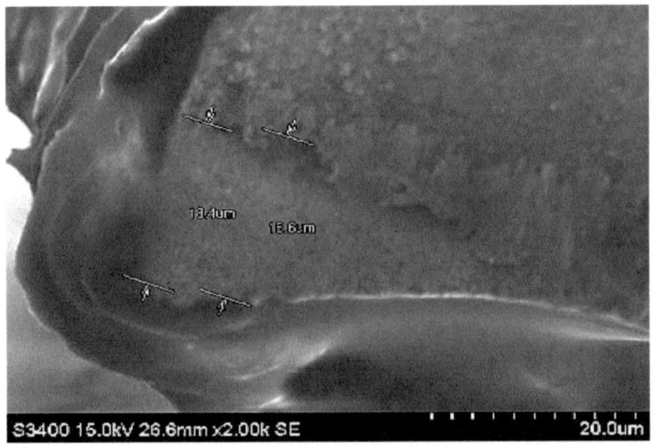

Figure 11.
Coated AlTiN tool wear at 70,000 rpm.

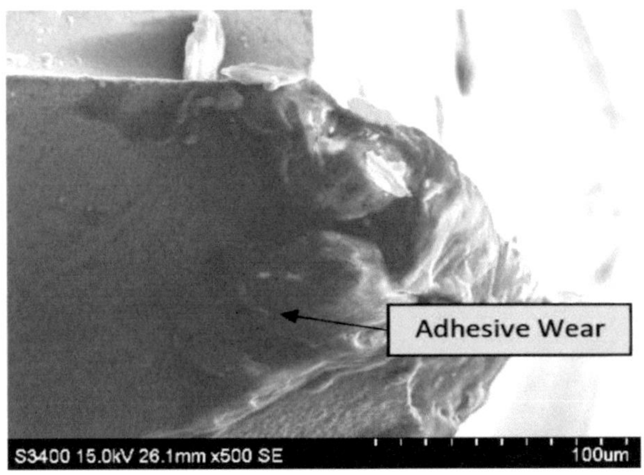

Figure 12.
Coated AlTiN tool wear at 110,000 rpm.

3.2 EDS results

The observed elemental composition on the worn surface of uncoated, coated TiAlN and AlTiN WC tools while machining the Ti-6Al-4V ELI titanium alloy (Grade 23) are shown in **Tables 4–6**.

SEM and EDS analysis for the considered machining operating parameters indicate the built-up edge, built-up layer and craters appearing at rake face and flank face of the cutting tool representing the adhesive, diffusion, abrasive and oxidation wear phenomenon on the tool surfaces. Wear confirmation process on the tool materials was discussed below.

3.2.1. SEM observations

SEM examination of the leading cutting edge for coated and uncoated tools indicates plastic deformation, adhesion wear, chipping and flaking as shown in **Figures 7–12**. In addition, coating delamination at the tool cutting edge and significant changes in the tool shape was observed under the SEM. Tool material reacts with titanium alloy by means of high chemical reaction, forming adhesive wear. During the tool-workpiece interaction, adhesive layers of tool material will break down that progresses to adhesion wear. The intimate contact between the tool and

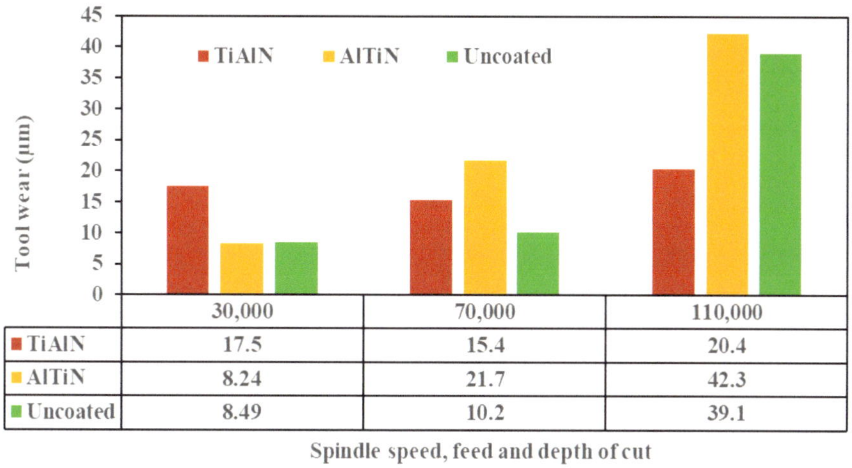

Figure 13.
Tool wear versus different machining conditions.

Maximum composition of elements	wt%
C	3–5.4
N	3–11.8
O	3–4.1
Al	0.2–0.7
Ti	0.3–10.2
V	0.3–1.1
Cr	0.3–0.8
Ba	2–9.8
W	70–84.5

Table 4.
EDS analysis results for uncoated tools.

Maximum composition of elements	wt%
C	6.8
N	9.8
O	9.4
Al	21.4
Ti	36.4
V	0.6
Cr	0.3–0.8
Ba	13.8
W	1.3

Table 5.
EDS analysis results for coated TiAlN tools.

chip interface leads to the friction and high-temperature generation, stipulating the transmission of atoms from tool material losing its hardness and ultimately breakdown happens. It appears that both coated and uncoated tool materials were

Maximum composition of elements	wt%
C	10.5
N	11.8
O	4.6
Al	23.6
Ti	41.1
V	0.5
Cr	0.4
Ba	13.7
W	1.9

Table 6.
EDS analysis results for coated AlTiN tools.

subjected to thermal and mechanical loads and could not be able to resist the wear during the interrupted cutting in the end milling process. The chip shape, segmented or continuous, decides the cutting temperature formation inciting to thermoplastic shear localization at the contact length, resulting in the diffusion process. The chip constituents and the rate of diffusion are controlled by cutting temperature. In the machining of titanium alloys, the machining parameters, particularly, cutting speed, influence the cutting temperature origination at the tool edge for initiating the diffusion process. The earlier researchers verified that cutting speeds generate high cutting temperature, a short contact length, a low shear angle and a high cutting pressure. Chip segmentation and tribological parameters—the physical medium—may be causing the coating delamination for both the coated tools while machining the titanium alloy. Since the thermal conductivities of the coating constituents are different, the heat flux q flows through the coating layer and penetrates the tool substrate. The reason behind the diffusion process at the tool substrate surface might be due to chemically instable Co binder elements. Adhesive surface between the coating layer and tool substrate surface was completely eliminated gradually by diffusion process as depicted in **Figures 7–12** and confirmed through EDS analysis.

3.2.2 EDS analysis

Adhesive, diffusion, abrasive and oxidation wear were the major means on the flank face. The elements observed on the cutting tool edge and workpiece indicate the diffusion process has taken place. From the EDS and SEM analyses, the existence of workpiece material constituents V, Al and Ti on the rake face wear land with built-up layer of cutting edge indicates diffusion might take place. Increasing cutting speed leads to the decrease in presence of V and Al at the tool cutting tool tip and it might be imaginable that Ti only sticks to the cutting tool edge. It has therefore been considered reasonable to suggest that the built-up layer was started through the sticking of Ti by different bonding actions, that is, directly proportional with temperature. Under very high speed cutting conditions, tool life depends directly on the formation of crater wear. Thin titanium oxide layer formation was observed on cutting edge of both the tools. In high-speed cutting conditions, cratering becomes so severe that the tool edge is weakened and eventually fracture which has been observed for the AlTiN and TiAlN tools at 110,000 rpm spindle speed, 0.1 mm depth of cut and 8 μm/tooth feed rate. At higher and lower spindle speeds, existence of cobalt (Co) was very negligible. This indicates that Co diffusion might take place from the tool material constituents allowing it to wear easily. Through the EDS analysis, the amount of carbon presence indicates its transfer from cutting tool material into the workpiece material, that is, diffusion process might be taking place. This transfer of carbon reacts with titanium forming TiC layer which is continuous at higher speeds known as chemical wear process, leading to crater formation on tool material as observed by earlier researchers. The occurrence of crater wear might be due to chip-tool contact stresses generation, depleting the C and Co from cutting tool. An adherent layer of TiC formation exists on the cutting edge due to the chemical reaction between the titanium workpiece and the cutting tool material. Formation of oxycarbides on the surface indicates the existence of oxygen. Earlier researchers suggested this existence by Auger spectroscopy Analysis. Researchers also suggested that TiC grains removal might be taking place from the cutting tool because of reduction in toughness as the diffusion process takes place. Replenishing of TiC grains on the tool surface will probably occur by obtaining C from WC grains [3, 12–17, 21–29, 32]. The presence of Mo, Ni, Br, Cr, Fe and V indicates the work material composition diffusivity to the tool tip. Sulphur (S), silicon (Si) and magnesium (Mg) act as protective layer or barrier for adhesive wear and prevent the welding and stiffening of the work material in the tool surface. Through EDS analysis, it can be predicted that coating delamination in the initial stages may be due to mechanical wear later on by chemical mechanisms.

4. Conclusions

Tool wear analysis of PVD-coated TiAlN and AlTiN and uncoated tungsten carbide tools in high-speed micro-end milling of alpha + beta Ti-6Al-4V ELI titanium alloy (Grade 23) was investigated by tool wear mechanisms formation using SEM and EDS analysis and cutting force analysis. If the spindle speed is maintained at constant rpm while increasing feed rate and depth of cut, then tool wear increases dramatically. By increasing spindle speed from 30,000 to 70,000 rpm while varying feed rate and depth of cut, then (i) tool wear remains constant for uncoated tools, (ii) tool wear increases for AlTiN-coated tools and (iii) tool wear remains constant for TiAlN-coated tools. By increasing spindle speed from 70,000 to 110,000 rpm while varying feed rate and depth of cut, then tool wear increases for all the tools. If coated TiAlN and coated AlTiN tools are compared, then TiAlN tools performed better for machining this alloy. Based on the investigations, it can be suggested that PVD-coated TiAlN tungsten carbide tools give better performance than PVD-coated AlTiN and uncoated tungsten carbide tools in HSMEM at 110,000 rpm when machining alpha + beta Ti-6Al-4V ELI titanium alloy (Grade 23). SEM and EDS analyses for the considered machining operating parameters indicate the built-up edge, built-up layer and craters appearing at rake face and flank face of the cutting edge representing the adhesive, diffusion, oxidation and abrasive wear phenomenon on the tool surfaces when machined on both titanium alloys. Based on the investigations, the tool and work material properties, feed rate, depth of cut and cutting speed influence the tool wear.

Acknowledgements

The authors gratefully acknowledge the support offered by Professor Dr. Ramesh Kumar Singh, Machine Tools Lab, Mechanical Engineering Department, IIT Mumbai, Maharashtra, India, in providing all the facilities for conducting the research work. This research program is financially supported under Approval Note No.368. MED/Institute Annual Grant provided by the Mechanical Engineering Department, SVNIT, Surat, Gujarat, India. The authors thank Technician Mr. Sagar Jagtap at Sophisticated Instrumentation Centre at SVNIT, Surat.

Conflict of interest

I would like to undertake that the abovementioned manuscript has not been published elsewhere, accepted for publication elsewhere or under editorial review for publication elsewhere.

Author details

Chakradhar Bandapalli[1*], Bharatkumar Mohanbhai Sutaria[2] and Dhananjay Vishnu Prasad Bhatt[2]

1 Department of Mechanical Engineering, Madanapalle Institute of Technology and Science, Madanapalle, Chittoor, Andhra Pradesh, India

2 Department of Mechanical Engineering, Sardar Vallabhbhai National Institute of Technology, Surat, Gujarat, India

*Address all correspondence to: chakri.b1@gmail.com

References

[1] Christoph L, Manfred P. Titanium and Titanium Alloys: Fundamentals and Applications. Germany: John Wiley & Sons; 2006

[2] Astakhov Viktor P. Tribology of Metal Cutting. London: Elsevier; 2006

[3] Mark Jackson J, Ahmed W. Surface Engineered Surgical Tools and Medical Devices. Springer Science: US; 2007

[4] Sha W, Malinov S. Titanium Alloys: Modeling of Microstructure, Properties and Applications. UK: Woodhead Publishing; 2009

[5] Elias CN, Lima JHC, Valiev R, Meyers MA. Biomedical applications of titanium and its alloys. Journal of the Minerals, Metals & Materials Society. 2008;**60**(3):46-49. DOI: 10.1007/s11837-008-0031-1

[6] Kai C. Machining Dynamics—Fundamentals, Applications and Practices. 1st ed. London: Springer; 2009

[7] Kai C, Dehong H. Micro-Cutting: Fundamentals and Applications. 1st ed. UK: Wiley; 2013

[8] Altintas Y. Manufacturing Automation-Metal Cutting Mechanics, Machine Tool Vibrations & CNC Design. 2nd ed. New York: Cambridge University Press; 2012

[9] Davim PJ. Machining of Titanium Alloys. New York: Springer; 2014

[10] Komanduri R, Reed WR. Evaluation of carbide grades and a new cutting geometry for machining titanium alloys. Journal of Wear. 1983;**92**(1):113-123. DOI: 10.1016/0043-1648(83)90011-X

[11] Kitagawa T, Kubo A, Maekawa K. Temperature and wear of cutting tools in high speed machining of Incone1718 and Ti-6A1-6V-2Sn. Journal of Wear. 1997;**202**:142-148. DOI: 10.1016/S0043-1648(96)07255-9

[12] Jawaid A, Sharif S, Koksal S. Evaluation of wear mechanisms of coated carbide tools when face milling titanium alloy. Journal of Materials Processing Technology. 2000;**99**(1):266-274. DOI: 10.1016/S0924-0136(99)00438-0

[13] Liu H, Sun Y, Geng Y, Shan D. Experimental research of milling force and surface quality for TC4 titanium alloy of micro-milling. International Journal of Advanced Manufacturing Technology. 2005;**79**(1-4):705-716. DOI: 10.1007/s00170-015-6844-5

[14] Nouari M, Ginting M. Wear characteristics and performance of multi-layer CVD-coated alloyed carbide tool in dry end milling of titanium alloy. Surface and Coatings Technology. 2006;**200**(18-19):5663-5676. DOI: 10.1016/j.surfcoat.2005.07.063

[15] Nouari M, Abdel-Aal HA, El Mansori M. Analysis of coating delamination under extreme contact loading. Tribology Letters. 2006;**23**(1):39-45

[16] Rahman M, Wang Z-G, Wong Y-S. A review on high speed machining of titanium alloys. JSME International Journal Series C. 2006;**49**(1):11-20. DOI: 10.1299/jsmec.49.11

[17] Ginta LT, Amin AKMN, Karim ANM, Patwari UA. Modeling and optimization of tool life and surface roughness for end milling titanium alloy Ti–6Al–4V using uncoated WC-co inserts. In: Proceedings of the International Conference at Curtin University of Technology Science and Engineering; 24-27 November 2008; Miri, Sarawak, Malaysia. 2008. pp. 24-27

[18] Schueler GM, Engmann J, Marx T, Haberland R, Aurich JC. Burr formation and surface characteristics in micro-end milling. In: Proceedings of the CIRP International Conference on Burr-Analysis, Control and Removal at University of Kaiserslautern; 2-3 April 2009; Germany. 2009. pp. 129-138. DOI: 10.1007/978-3-642-00568-8

[19] Arrazola PJ, Garay A, Iriarte LM, Armendia M, Marya S, Maitre FL. Machinability of titanium alloys (Ti6Al4V and Ti555.3). Journal of Materials Processing Technology. 2009;**209**(5):2223-2230. DOI: 10.1016/j.jmatprotec.2008.06.020

[20] Malekian M, Park SS, Jun MBG. Tool wear monitoring of micro-milling operations. Journal of Materials Processing Technology. 2009;**209**(10):4903-4914. DOI: 10.1016/j.jmatprotec.2009.01.013

[21] Smith PWB, Axinte DA, Limvachirakom V. Preliminary study of the effects of crystal orientation of a CVD monocrystalline diamond in micromilling of Ti–6Al–4V. Journal of Engineering Manufacture. 2010;**224**(8):1305-1312. DOI: 10.1243/09544054JEM1861SC

[22] Zhang S, Li JF, Sun J, Jiang F. Tool wear and cutting forces variation in high speed end-milling Ti-6Al-4V alloy. International Journal of Advanced Manufacturing Technology. 2010;**46**: 69-78. DOI: 10.1007/s00170-009-2077-9

[23] Ozel T, Thepsonthi T, Ulutan D, Kaftanolu B. Experiments and finite element simulations on micro-milling of Ti-6Al-4V alloy with uncoated and CBN coated micro-tools. CIRP Annals—Manufacturing Technology. 2011;**60**(1):85-88. DOI: 10.1016/j.cirp.2011.03.087

[24] Thepsonthi T, Ozel T. Experimental and finite element simulation based investigations on micro-milling Ti-6Al-4V titanium alloy: Effects of CBN coating on tool wear. Journal of Materials Processing Technology. 2013;**213**(4):532-542. DOI: 10.1007/s00170-012-3980-z

[25] Thepsonthi T, Ozel T. Multi-objective process optimization for micro-end milling of Ti-6Al-4V titanium alloy. International Journal of Advanced Manufacturing Technology. 2012;**63**(9-12):903-914

[26] Wyen CF, Jaeger D, Wegener K. Influence of cutting edge radius on surface integrity and burr formation in milling titanium. International Journal of Advanced Manufacturing Technology. 2013;**67**(1-4):589-599. DOI: 10.1007/s00170-012-4507-3

[27] Durul U, Ozel T. Machining induced surface integrity in titanium and nickel alloys: A review. International Journal of Machine Tools and Manufacture. 2011;**51**(3):250-280. DOI: 10.1016/j.ijmachtools.2010.11.003

[28] Bajpai V, Kushwaha AK, Singh RK. Burr formation and surface quality in high speed micromilling of titanium alloy (Ti-6Al-4V). In: ASME International Manufacturing Science and Engineering Conference on Micro and Nano Technologies; Sustainable Manufacturing; 10-14 June 2013; Madison, Wisconsin, USA. Vol. 2. 2013. pp. 1216-1224. DOI: 10.1115/MSEC2013-1216

[29] Hou J, Zhou W, Duan H, Yang G, Xu H, Zhao N. Influence of cutting speed on cutting force, flank temperature, and tool wear in end milling of Ti-6Al-4V alloy. International Journal of Advanced Manufacturing Technology. 2014;**70**:1835-1845. DOI: 10.1007/s00170-013-5433-8

[30] Kim DH, Lee P-H, Lee SW. Experimental study on machinability of Ti-6Al-4V in micro end milling. In: Proceeding of the World Congress on

Engineering; 2-4 July 2014; London, UK. 2014

[31] Pervaiz S, Rashid A, Deiab I, Nicolescu M. Influence of tool materials on machinability of titanium and nickel based alloys: A review. Journal of Materials and Manufacturing Processes. 2014;**29**:219-252. DOI: 10.1080/10426914.2014.880460

[32] Hassanpour H, Sadeghi MH, Rezaei H, Rasti A. Experimental study of cutting force, micro hardness, surface roughness, and Burr size on micromilling of Ti6Al4V in minimum quantity lubrication. Journal of Materials and Manufacturing Processes. 2016;**31**(13):1654-1662. DOI: 10.1080/10426914.2015.1117629

[33] Bandapalli C, Singh KK, Sutaria BM, Bhatt DV. Experimental investigation of machinability parameters in high speed micro-end milling of titanium (Grade-2). International Journal of Advanced Manufacturing Technology. 2016;**85**:2139-2153. DOI: 10.1007/s00170-015-7443-1

[34] Jaharah Ghani A, Haron CHC, Kasim MS, Sulaiman MA, Tomadi SH. Wear mechanism of coated and uncoated carbide cutting tool in machining process. Journal of Materials Research. 2016;**31**(13):1873-1879. DOI: 10.1557/jmr.2015.382